ALSO BY AMIR D. ACZEL

Entanglement: The Greatest Mystery in Physics

*The Mystery of the Aleph: Mathematics, Kabbalah,
and the Search for Infinity*

*Fermat's Last Theorem: Unlocking the Secret of
an Ancient Mathematical Problem*

*Chance: A Guide to Gambling, Love, the Stock Market,
and Just About Everything Else*

*The Riddle of the Compass: The Invention
That Changed the World*

Pendulum: Léon Foucault and the Triumph of Science

*God's Equation: Einstein, Relativity,
and the Expanding Universe*

angulis plani[...] duceretur per minimum angulos[...]
anguli ecchi plani, remaneret aggregatum[...]
corporis p[...]

[...] media pars numeri angulorum solidi[...]
[...] pyramides duplicati[...] media pars numeri[...]
angulorum punctis plura corpora, in quolibet[...]
[...] ad minimum triplo plures anguli[...]
binarius[...] numerus angulorum solidorum[...]
ducatur[...] binarium fit, usque[...]
[...] binarium, si quadrum sit numerus[...]
[...] per plana quaeres[...] addendo binario[...]
[...] inter facies & angulos/solidos[...]
describuntur. [...] rectanguli[...]

[...] demonstrat[...] plura[...]
pro numero angulorum solidorum[...]

$$\frac{2\alpha - 4}{1\alpha} \quad \text{et} \quad \frac{2z - 4}{1\alpha}$$

potest, si a sit 4. 6. 8. 12. 20.
[...] corpora regularia[...]
[...] omnes[...]
[...]
[...] in centro sph[...]
a centro sphaero[...]
qui simul concurrant cum illis[...]
Dato aggregato ex omnibus angulis planis[...]
invenire[...] in eodem corpore solido[...]
et per id idem, invidiatur per[...] residuum[...]
et nullum tale corpus esse potest.
[...] faciendum numerum angulorum[...]
et productum[...] addatur[...]
[...] media pars ejus numeri angulorum[...]
angulis planis est[...] numeri faciemus 12. cujus[...]
120 cujus media pars est 60. ergo in tali[...]
sunt semper duplo plures anguli plani[...]
[...] latus semper commune est duobus[...]
numerum planorum, [...]

DESCARTES'
Secret Notebook

A TRUE TALE OF MATHEMATICS, MYSTICISM, AND
THE QUEST TO UNDERSTAND THE UNIVERSE

AMIR D. ACZEL

Broadway Books
New York

PUBLISHED BY BROADWAY BOOKS

Copyright © 2005 by Amir D. Aczel

All Rights Reserved

A harcover edition of this book was originally published in 2005 by Broadway Books.

Published in the United States by Broadway Books, an imprint of The Doubleday Broadway Publishing Group, a division of Random House, Inc., New York. www.broadwaybooks.com.

BROADWAY BOOKS and its logo, a letter B bisected on the diagonal, are trademarks of Random House, Inc.

Illustrations by Kymtra Design

Library of Congress Cataloging-in-Publication Data
Aczel, Amir D.
Descartes' secret notebook : a true tale of mathematics, mysticism, and the quest to understand the universe / Amir D. Aczel.
p.   cm.
1. Descartes, René, 1596–1650—Notebooks, sketchbooks, etc. 2. Descartes, René, 1596–1650—Knowledge—Mathematics. 3. Mathematics—Philosophy—History—17th century. 4. Leibniz, Gottfried Wilhelm, Freiherr von, 1646–1716—Manuscripts. I. Title.
QA29.D55A25 2005
510'.1—dc22
2005045722

ISBN-13: 978-0-7679-2034-6
ISBN-10: 0-7679-2034-1

PRINTED IN THE UNITED STATES OF AMERICA

10   9   8   7   6   5   4

For Debra

ACKNOWLEDGMENTS

I am extremely grateful to the John Simon Guggenheim Memorial Foundation for making me a Fellow of the Foundation. My Guggenheim Fellowship made this book possible, and receiving this award has been the greatest honor in my professional career. I thank the foundation and its officers for believing in me and in this project even before its acceptance by a publisher. My special thanks to Senior Vice President G. Thomas Tanselle for his interest in my work.

I thank the librarians of the Bibliothèque nationale de France in Paris for giving me access to many original manuscripts and documents on Descartes and the mystery of his notebook. Also in Paris, I am indebted to the Institut de France and its librarians for allowing me to use the institute's extensive archives, and to Professor Jean Dercourt, Perpetual Secretary of the French Academy of Sciences.

Many thanks to the Gottfried Wilhelm Leibniz Library in Hanover, Germany, and to its administrator, Birgit Zimny, for making available to me the manuscripts Leibniz copied from Descartes, which held the key to the mystery addressed in this book. My appreciation for the photographic processing work on the Leibniz images performed by Kevin Wool and Boston Photo Imaging.

I am grateful to Professor Jay M. Pasachoff of Williams College for providing me with his image of Kepler's model of the universe, and I thank Wayne G. Hammond of the Chapin Library at Williams College for making this figure available to me.

I want to express my many thanks to Jeff Weeks (www.geometry games.org) for providing me with his images of the Poincaré dodecahedral space and other geometrical models of the universe, and for explaining to me his work in cosmology.

I thank the Descartes Museum in Descartes, Touraine, and its administrator, Daisy Esposito, for guiding me to a number of important documents on the early life of René Descartes. I thank the Protestant Museum in La Rochelle, France, for information on the siege of the city in the seventeenth century.

I am grateful to the photographer Tzeli Hadjidimitriou (www. odoiporikon.com) in Athens, Greece, for her photograph of the temple on the island of Delos.

I wish to express my gratitude to Professor Owen Gingerich and the department of the history of science at Harvard University for appointing me a visiting scholar at the department.

I acknowledge the Center for the Philosophy and History of Science at Boston University, where I spent a year as a Research Fellow while writing this book. My work was aided by a number of people at the center, including Alfred Tauber and Debra Daugherty, and at various branches of the Boston University Library.

My deepest gratitude to my agent and friend, John Taylor (Ike) Williams, of Kneerim and Williams in Boston, for guiding my writing career with so much patience and wisdom.

My heartfelt thanks also to Hope Denekamp of Kneerim and Williams for all her tireless help with this book, and with everything related to publishing.

I am very grateful to Gerald Howard, my editor at Broadway Books in New York, for his clear judgment, knowledge, and guidance in shaping a manuscript into a book, and for always keeping me on the right track in this complex endeavor. I thank Rakesh Satyal for all his work on this project, for his boundless energy, and for his care in making this project a reality.

Many colleagues and friends helped me in the undertaking of researching and writing this book. They include Judith Alvarez-Perreire, Dan Carey, Stephen Gaukroger, Kurt Hawlitschek, Richard Landes, Kenneth Manders, Michael Matthews, Jacob Meskin, Evelyne Patlagean, Arthur Steinberg, and Marina Ville. Thank you, everyone.

My deepest thanks and appreciation to my wife, Debra, for helping me edit and revise the manuscript, for photographic work, and for her many important ideas. This book is dedicated to her.

Contents

Introduction

I HELD THE FRAGILE ANCIENT MAN-
uscript in my hand. I opened it carefully,
and read:

PREAMBLES

Fear of God is the beginning of
wisdom. The actors, called to the
scene, in order to hide their flaming cheeks, don a mask. Like them,
when climbing on stage in the
theater of the world, where, thus far,
I have only been a spectator, I advance masked. At the time of my
youth, witnessing ingenious discoveries, I asked myself whether I could
invent all on my own, without lean-
ing on the work of others. Henceforth, little by little, I became aware
that I was proceeding according to

determined rules. Science is like a woman: if faithful, she stays by her husband, she is honored; if she gives herself to everyone, she is degraded.

The manuscript continued further. After a few more pages, I read another fragment of text:

OLYMPICA

November 11, 1620. I began to conceive the foundation of an admirable discovery.

These were the enigmatic words of René Descartes (1596–1650). They were never intended for eyes other than his own. But the manuscript I now held in my hands was not written by Descartes. It was a copy of Descartes' secret writings made by none other than Gottfried Wilhelm Leibniz (1646–1716)—one of the greatest mathematicians of all time and the man who, only a few years after copying Descartes' notebook in Paris in 1676, would give us the calculus.

The idea for a book about Descartes occurred to me while lost in a snowstorm somewhere in eastern Ontario, hard by the Quebec border, around midnight in early January of 2002. We were on our way back home to Massachusetts from a holiday visit to relatives in Toronto, hoping to spend the last night of our trip in Montreal, when our car got mired in deep snow dropped by a sudden blizzard. I turned off the highway to look for a place to wait out the storm; but after making an unfortunate sequence of turns onto various country roads, I came to the conclusion that I had gotten us completely lost. The visibility was poor, there were no lights in sight, and we had no idea where to turn. I knew that if the fuel ran out, eventually we would start to freeze.

While driving, I glanced at my dashboard. And I remembered something this car came equipped with—something I had never had the need to use before. On my dashboard, I recognized the small lit button, and I pressed it. There followed the sound of a telephone number being dialed automatically. "Good evening, Mr. Aczel," said a pleasant voice from a thousand miles away. "How are you this evening? I see that you are heading south on Glen Donald Road, half a mile north of the intersection with County Road 27, outside of Cornwall, Ontario."

"Aha . . . ," I said, trying to keep my voice from betraying that I had absolutely no idea where we were. "We are looking for a hotel. . . . It's snowing hard."

"No problem at all," was the response. "Where are you trying to go?"

"Montreal."

"That's not hard," reassured the voice from civilization. "Make a left turn at the next intersection, the one with County Road 27. Follow that road for two miles, and then turn right at the intersection, and you will see the on-ramp for Highway 401 to Montreal. Which hotel would you prefer? There's a Marriott just as you reach the city from this direction. I can make you a reservation, and I'll be back with more directions as you continue."

The voice on this cell phone led us throughout the night: from a deserted country road in snowy Ontario all the way to a warm hotel room in Montreal, directing us at every turn along the way. The person giving us instructions from afar knew our position, accurate to within a dozen feet, during every minute of the trip. The technology that made all this possible is the Global Positioning System (GPS), which allows one to determine the location of a small radio receiver (called a GPS receiver) anywhere in the world. In this case, a device installed in my car, linked to an internal cellular phone, allowed the company that provides this service to determine the location of my vehicle wherever it might be. The amazing GPS technology works because of an invention made almost four centuries ago by the philosopher, scientist, and mathematician René Descartes.

Descartes gave us the *Cartesian coordinate system*, named after him—a system of crisscrossed parallel lines, in two, three, or more dimensions, that allows us to *describe numerically* the position of a point in space. In this case, the position of the point (my car) was described in terms of its latitude and longitude, which were then translated into a location on a map. The GPS system works in three dimensions as well: it can give you, in addition to your latitude and longitude, your altitude—and this makes it useful in directing an airplane.

But Descartes' coordinate system is used for a lot more than GPS. Every pixel on your computer screen is described internally by a pair of numbers: its horizontal and vertical coordinates. Thus all computer technology relies on Descartes' invention. Graphs and diagrams and maps of all kinds rely on the Cartesian system, and so do digital photographs, so popular these days, and pictures and documents sent on the Internet, and engineering designs, and space flights, and oil exploration.

And the applications go even further—beyond our three-dimensional intuition of pictures and graphs and shapes. When data are available on many variables—more than the visual three dimensions of everyday life—such data can *still* be analyzed using the Cartesian coordinates. Your bank, for example, may have data on your income, your assets, the number of years at your current employment, the number of people in your family, your age, your educational level, and so forth. These *multidimensional* data can be mapped on a many-variable scale using the Cartesian coordinates (even though such a "map" cannot be visualized and is meaningful only within the context of the computer analyzing the data) and through statistical analysis lead to a decision by the bank to approve or deny your loan application. Statistical and scientific algorithms that analyze data on many variables use the Cartesian system in the analysis. The number of applications of the Cartesian coordinate system in our daily lives is immense. Literally everything we do or see or use in our daily lives has something to do with Descartes' great invention.

Interestingly, Descartes' invention of the coordinate system that bears his name was a special outcome of a much grander design. Descartes achieved an immense advance in mathematics, launching modern mathematical theory four centuries ago, when he unified algebra with geometry by inventing analytic geometry: a way of connecting the equations and formulas of algebra with the figures and shapes of geometry. The Cartesian coordinate system was just the device he created in order to facilitate this unification.

Of course, Descartes' greatest fame comes not simply from his work in mathematics or in physics—in which he also made important discoveries, especially on gravity and falling objects, as well as in optics—but from his philosophy. Descartes' *"Cogito, ergo sum"* (I think, therefore I am), and the philosophy behind this statement, is a pillar of modern philosophy; and his rationalism—*Cartesianism*—is considered of great importance in the development of philosophical thinking. Descartes is often viewed as the founder of modern philosophy. In his *Meditations*, published in 1641, Descartes wrote: "It must be acknowledged that this pronouncement, *I am, I exist*, whenever I assert it or conceive it in my mind, is necessarily true." In the introduction to the book they edited, *Descartes and His Contemporaries*, M. Grene and R. Ariew describe Descartes' statement above as signaling a turning point in Western thought: "Suddenly, we have reached a new level of awareness, at which we could ask reflective questions about ourselves: Can we reach out from consciousness to an external world? Who are we as minds in relation to our bodies?" Some scholars have even asserted that Descartes' philosophy, in its introduction of the self into human consciousness, inaugurated modern psychological theory. Descartes' method of reasoning thus created self-reflection, which incorporated into philosophy the elements of modern psychology. Descartes pioneered metaphysical investigations, and hypothesized about the relation between body and soul. He tried to use reason and logic to prove the existence of God (in whom he believed).

There is a direct link between Descartes' logical, rational approach to philosophy and his work in mathematics. The reason for this is that Descartes' philosophy is based on an ambitious attempt to found all of human knowledge on the same precise, strictly logical principles that the ancient Greeks had used in creating their enduring geometry.

I believe that Descartes' work in mathematics, physics, and philosophy, as well as in other areas this unique individual studied, such as biology, anatomy, and music theory, are unified by invisible links of logic. This recondite internal rationality is what Descartes was all about—for the hidden Descartes was fundamentally a supreme mathematician: a man who was so good at doing mathematics that he came to believe he could apply his mathematical skills and methods to every area of human study.

Descartes lived in one of the most tumultuous, and yet intellectually fecund, periods in history. Descartes' time, the first half of the seventeenth century, was the era of the Thirty Years War, in which Catholics and Protestants were mercilessly pitted against each other in a series of bloody battles. This period also saw the ruthless suppression by the Catholic Church of new scientific and philosophical ideas, as evidenced by the trial of Galileo by the Inquisition, the persecution by the church of other thinkers who supported the theory of Copernicus, and the burning of forbidden books. However, this was also a period of great intellectual revival—an extension of the Renaissance to science, mathematics, and philosophy. Classical ideas in these areas were being studied and extended by intellectuals throughout Europe. Descartes' work was both a product of this period and the vanguard, leading the way in the development of mathematics and philosophy right up to our time.

My favorite café in all Paris is Les Deux Magots—that icon of literary society made famous by Hemingway, Fitzgerald, Sartre, and Beauvoir— facing the ancient church of Saint-Germain-des-Prés. Six months after our Canadian trip in the snow, I found myself on a sunny day sitting at this café with my friend the historian Richard Landes. We were drinking iced coffee, and Richard was telling me about arrangements he wanted to make for me to meet the director of the Descartes center in Paris, now that I'd come to the French capital to research the life of the philosopher and mathematician.

I was renting an apartment in the Marais—the oldest, medieval part of Paris, and the only section of the city that had not been razed in the nineteenth century by Baron Georges Haussmann as part of his plan to build the wide and elegant boulevards we associate with Paris today. This part of the city, with its narrow ancient streets, looks very much as it appeared in the days of Descartes. My apartment, on the rue du Bourg-Tibourg, was in a building erected in 1629. Both the building and the apartment, on which apparently little work had been done in the past few centuries other than necessary maintenance, looked very much as they must have looked when Descartes walked these very streets when he lived just around the corner from this address for a short time in 1644: on the rue des Ecouffes, between the rue du Roi de Sicile and the rue des Blancs-Manteaux.

The apartment consisted of one large room with a kitchen corner and a bath, and it had a high ceiling supported by the original dark brown wooden beams so characteristic of seventeenth-century French construction. A narrow, dark spiral staircase with small windows cut into the thick stone walls led up to the apartment. Looking at the building from outside, one could admire the tall Parisian windows and notice the protruding iron supports holding together the exterior walls of this ancient structure. Knowing that my building was there at the time Descartes walked these same streets lent my search an added sense of reality.

I spent my days at the libraries and archives of Paris, researching material on Descartes and his work; I traveled to the locations at which he stayed or lived throughout Europe—Descartes was a great traveler who saw most of the Continent—and I walked the perimeter of the Place des Vosges, under the ancient arches, exactly as Descartes did in 1647 discussing mathematics with Blaise Pascal. I held in my hands original letters Descartes had written to his friend Marin Mersenne; I perused countless manuscripts written over the centuries; I even bought an original book of Descartes', published in 1664. But sometime in the middle of my search, I made a surprising discovery: Descartes had kept a secret notebook.

I was now sitting at the heart of the area in which Descartes loved most to live whenever he was in Paris for longer stays—the then-as-now-fashionable district of Saint-Germain-des-Prés. Richard was talking fast about Descartes and history and Paris, but we were too often interrupted by the perpetual ring of his cell phone. I looked at the ancient church in front of us. I knew it was very old—its construction began in the sixth century. The church has a graceful tower, dating to the tenth century and still in its original state. It has a rustic kind of beauty, seen more often in churches in the French countryside, and in fact it used to be out in the fields, outside the city walls—hence the designation "des Prés." And I knew something else about this church: inside it is a crypt containing the remains of René Descartes. But the body of the great philosopher and mathematician—so revered by the French—lacks a head. Descartes' skull, or rather a skull purported to be that of the philosopher, is displayed elsewhere in Paris. Nothing about Descartes is simple, and nothing is what it seems, as I learned in my quest to understand Descartes and to uncover his secrets.

Leibniz's Search in Paris

ON JUNE 1, 1676, GOTTFRIED WIL-helm Leibniz, who would become known as one of the greatest mathematicians of all time and who—along with Newton, who worked independently in England—would be credited with the discovery of the calculus, stepped off a carriage in front of a house in Paris, climbed up a set of stairs, and knocked on a heavy wooden door.

Leibniz had arrived in Paris a few years earlier from Hanover, in his native Germany. He was on a diplomatic mission on behalf of his patron, a German noble-man. But personally, Leibniz was on a search for Descartes' hidden writings. He had heard that when Descartes died, in Stockholm in 1650, he had left behind a locked box containing writings that he never intended to publish and had kept secret throughout his life. Leibniz knew that these writings were kept somewhere in the French capital, and over three and a half years in Paris, Leibniz made every effort to locate this treasure. Finally, using a network of connections, Leibniz was able to obtain the name and address of Claude Clerselier (1614–84), a man who had been a friend of Descartes and an editor and translator of his works.

Leibniz learned that a quarter of a century earlier, Clerselier received Descartes' hidden manuscripts as a present from his brother-in-law Pierre Chanut (1601–62), who had been France's ambassador to Sweden and a confidant of Descartes during the few months he served as philosophy teacher to Queen Christina of Sweden before he died.

Some time after Descartes' death, Ambassador Chanut loaded the box containing Descartes' hidden manuscripts onto a ship in Stockholm headed for France. After long delays en route, the cargo was unloaded sometime in 1653 at the French port of Rouen. The box was then reloaded onto a boat that was to take it up the Seine River to Paris. But just as the boat entered Paris and was passing by the palace of the Louvre, it capsized and sank. The sealed box containing Descartes' manuscripts remained submerged for three days. Then, miraculously, it disentangled itself from the wreckage and was found on the bank of the river some distance downstream.

Upon hearing this news, Clerselier—who had been expecting the precious cargo for a long time and had sadly just given up hope of ever seeing the manuscripts after he heard that the boat had capsized—rushed to the river with all his servants and ordered them to retrieve the papers quickly. He instructed his servants to spread the parchment sheets of Descartes' manuscripts on tables in his house to dry. The servants were illiterate, and found it difficult to reassemble the manuscripts. But Clerselier worked hard to save Descartes' hidden writings, and spent many years reading the manuscripts and putting them in order. There was, however, one notebook whose content he could not understand.

The elderly man opened the door slightly, but seeing someone he did not recognize, shut it again. "Please," pleaded the young man from behind the locked door, "please read this letter," and he thrust in through a reopened crack in the door a letter of introduction from the duke of Hanover, asking whoever was shown it to offer all help to its bearer.

After quickly reading the letter, Clerselier opened his door and motioned Leibniz in. Clerselier was a possessive man, and had been jealously guarding Descartes' writings. He viewed himself as the protector of his late friend's secrets. Clerselier listened intently as Leibniz explained his urgent and unusual need to see the documents. Upon hearing the story, he understood that this young man's future and reputation might well depend on the content of Descartes' hidden writings. So reluctantly, and despite his inclination to the contrary, Clerselier agreed to let Leibniz see Descartes' work, and even to copy it.

Leibniz sat down, opened a manuscript, and read:

PREAMBLES

Fear of God is the beginning of wisdom. The actors, called to the scene, in order to hide their flaming cheeks, don a mask. . . .

But after reading Descartes' description of his hope of discovering science all on his own, and of "advancing masked" through life, Leibniz read the following as the manuscript continued:

Providing the reader with the true means of resolving all the difficulties of this science; it is demonstrated that, on these difficulties, the human spirit can find nothing more. This, to shake off idle chatter, and to dismiss the recklessness of some who promise to demonstrate in all the sciences new miracles.

Leibniz understood that Descartes had planned to write a book about an important mathematical discovery using a pseudonym. René Descartes would be Polybius the Cosmopolitan. Leibniz paused for a moment. Then he continued to peruse the unusual document; but what he read next startled him:

Offered, once again, to the erudite scholars of the entire world, and especially to G. F. R. C.

In the copy he made of the manuscript, Leibniz added a parenthetical word, making it read:

G. (Germania) F. R. C.

He did not need to annotate the acronym "F. R. C." He knew it well . . . perhaps too well. Leibniz felt a quiver run down his spine as he realized that an invisible secret bond linked him with the late French philosopher.

Looking carefully at the manuscripts in front of him, Leibniz understood that the *Preambles* and the *Olympica*, which announced Descartes' "admirable discovery" but did not give the discovery itself, were only fragments designed to introduce the actual work in which the truth Descartes did not disclose in these manuscripts was exposed. But what was the work, and where was it? As he was about to find out, Leibniz was now very close to Descartes' deepest secret, the discovery nearest to Descartes' heart—one that would need to be veiled with a pseudonym, mysterious language, and bizarre, mystical notation.

"Yes, there is one more item," said the elderly gentleman after Leibniz had finished five days of copying and eagerly asked him if there was anything else. "But nobody else has ever seen it before. It's a notebook—his secret notebook." Then he added, "But I don't think you would understand it anyway. I've worked on it for years, but nothing in it, symbols, drawings, formulas, makes any sense at all. It is completely coded."

Leibniz pleaded, and again explained his desperate need to learn everything he could about the hidden work of Descartes. He promised that he would keep the secret, whatever it was that was hidden in Descartes' pages. Finally, Clerselier relented, but imposed tight restrictions on the access to this notebook.

Descartes' notebook consisted of sixteen parchment pages. It contained bizarre notation. Some of the symbols resembled those associated with alchemy and astrology—not the characters usually found in writings on mathematics. And next to them were strange, obscure figures. Then there were seemingly incomprehensible sequences of numbers. What did all these mean?

Working intently and very fast—perhaps furtively, for we don't know exactly what were the conditions Clerselier had imposed when he finally allowed Leibniz to see the notebook—Leibniz had to decipher Descartes' code while at the same time doing the copying. He was able

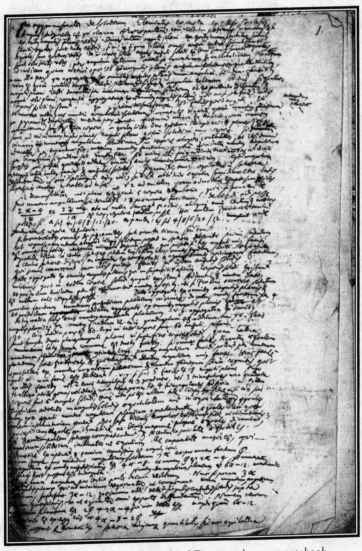

A page from Leibniz's copy of Descartes' secret notebook

to copy only one and a half pages by the time he had to stop his work. Part of Leibniz's copy of Descartes' secret notebook is shown on page 15.

Some years after Leibniz made this copy, Descartes' original notebook disappeared forever. And for over three centuries, no one could understand the meaning of the copy Leibniz had made of Descartes' secret notebook.

What the bizarre symbols, including $2\!\!\!/$, meant, and what the sequences of numbers—

$$4 \quad 6 \quad 8 \quad 12 \quad 20 \quad and \quad 4 \quad 8 \quad 6 \quad 20 \quad 12$$

—signified, remained a deep mystery.

Why did Descartes keep a secret notebook? What were its contents? And why did Leibniz feel compelled to travel to Paris, seek out Clerselier, and copy pages from Descartes' notebook?

Chapter 1

The Gardens of Touraine

JUST BEFORE RENÉ DESCARTES WAS born, on March 31, 1596, his mother, Jeanne Brochard, took an action that may well have altered the course of Western civilization. For like Julius Caesar crossing the Rubicon in 49 B.C., Jeanne Brochard crossed the Creuse River, which lay between her family home, in the region of Poitou, and the small town of La Haye, which lies in the region of Touraine, in western central France.

The Descartes family had originated in Poitou and had lived for many years in the town of Châtellerault, about twenty-five kilometers south of La Haye. Descartes' parents, Joachim Descartes and Jeanne Brochard, who were married on January 15, 1589, owned a stately man-

The Descartes family mansion in Châtellerault

sion in the center of Châtellerault, at 126 rue Carrou-Bernard (today's rue Bourbon).

Joachim Descartes was the councillor of the Parliament of Brittany, and this important job kept him away in distant Rennes. Jeanne needed her mother's help in birthing the baby, and this is why she traveled north and across the river to Touraine to give birth to René Descartes

Descartes' grandmother's house (now the Descartes Museum)

in her mother's house in La Haye. Sometime later, once she had recovered, she returned to Châtellerault. Despite this accident of birth, throughout his life, René's friends would often call him René *le Poitevin*—René of Poitou.

The regions of Poitou and Touraine include pastoral farmlands that have been cultivated since antiquity. There are low hills, many of which are forested, and rich flatlands, irrigated by rivers that cut through this fertile land. Cows and sheep graze here, and many kinds of crops are grown. La Haye is a small town of stone houses with gray roofs. At the time of Descartes, the population of the town numbered about 750 people.

Châtellerault is a larger, more genteel town than La Haye, with wide avenues and an elegant city square, and it serves as the hub of rural life in the region. Because this part of France is so fertile and rich in water and agricultural resources, the people who live here are well off. North of La Haye one can still visit the beautiful châteaus of the Loire Valley, as well as forests and game reserves, which existed at the time of Descartes. The châteaus, many of them restored to their original state, with lavish fifteenth- and sixteenth-century furnishings and surrounded by sculpted gardens, give us a feel for the life of the rich at the time of Descartes.

While the regions of Poitou and Touraine are similar in their topography, scenery, and the way the towns and villages are laid out, there was one important difference between them. While Poitou was mainly Protestant, Touraine was mostly Catholic. We know that in the fifteen years from 1576 to 1591, there were only seventy-two Protestant baptisms in La Haye. This significant religious difference between the two regions would affect the life of René Descartes. For this accident of birth—being born, and later also raised, in a strongly Catholic region while his family hailed from a Protestant one—would exert a significant impact on René's personality, thus influencing his actions throughout his life and determining the course of development of his philosophical and scientific ideas and the way he divulged them to the world.

Descartes lived in a century that knew severe tensions, including wars, between Catholics and Protestants. The fact that he was born in a Catholic region and would be raised by a devout Catholic governess, while many of his family's friends and associates in Poitou were Protestants, contributed to Descartes' natural secretiveness. It also made him, as an adult, much more concerned about the Catholic Inquisition than perhaps he should have been, and not worried enough about the persecution he could face from Protestants. Consequently, Descartes would refrain from publishing elements of his science and philosophy for fear of the Inquisition, and yet would readily settle in Protestant countries, where academics and theologians would viciously attack his work in part because they knew he was a Catholic.

René Descartes was baptized a Catholic in Saint George's Chapel in La Haye, a twelfth-century Norman church, on April 3, 1596. His baptismal certificate reads:

> This day was baptized René, son of the noble man Joachym Descartes, councillor of the King and his Parliament of Brittany, and of damsel Jeanne Brochard. His godparents, noble Michel Ferrand, councillor of the King, lieutenant general of Châtellerault, and noble René Brochard, councillor of the King, judge magistrate at Poitiers, and dame Jeanne Proust, wife of monsieur Sain, controller of weights and measures for the King at Châtellerault.
>
> **[signed]**
>
> **Ferrand   Jehanne Proust   René Brochard**

Michel Ferrand, René Descartes' godfather, was his great-uncle—Joachim Descartes' uncle on his mother's side. René Brochard was the baby's grandfather, Jeanne Brochard's father. Jeanne's mother, René Descartes' grandmother, was Jeanne Sain. Her brother's wife was Jeanne (Jehanne) Proust.

The name La Haye comes from the French word *haie*, meaning hedge. Originally the town was named Haia, and in the eleventh century, as the language evolved, the spelling was changed to Haya. In Descartes' time it was called La Haye–en–Touraine. In 1802, the town's name was changed to La Haye–Descartes, in honor of the philosopher, and in 1967, "La Haye" was altogether dropped and the town is now known as Descartes.

There was once a castle in the area, and the wealthiest and most powerful citizens of La Haye lived in it. As the feudal system disintegrated, the castle was abandoned and the people moved down to the town, where they could live more comfortably. But life in this area remained difficult. People suffered both from wars and from disease. The plague ravaged La Haye and its surroundings several times in the late sixteenth and early seventeenth centuries. In 1607, the town was placed under quarantine after a recurrence of the plague killed many people in the area.

In early times, the hedges denoted by the French name Haia (or Haya or Haie) stood for thorny hedges that were planted by the townspeople in an effort to defend themselves and their property from marauding bands of brigands and highwaymen that pillaged and looted the countryside. But as times improved after the duke of Anjou took control of the area in September 1596—six months after the birth of René Descartes—the hedges implied by the name of the town came to mean hedges of beautiful gardens. To this day, this part of France is known for its exquisite gardens and areas of great natural beauty.

Late in life, just before he left on his final journey to Stockholm, Descartes wrote to his dearest friend, Princess Elizabeth: "A man who was born in the gardens of Touraine, shouldn't he avoid going to live in the land of the bears, between rocks and ice?" Descartes may well have been thinking of the garden of his own childhood home, his grandmother's house in La Haye, when he wrote these words to Elizabeth. Descartes' childhood home is an attractive two-story country house with four large rooms, although it is not nearly as impressive as the family mansion in Châtellerault. But it is surrounded by an exuberant garden, now restored to its original state, with flowers under the generous canopy of graceful trees. One can imagine the young boy enjoying many hours of undisturbed thinking and playing in this tranquil garden.

A year after René was born, shortly after giving birth to her fourth child, Jeanne Brochard died. The newborn survived for three more days, and then died too. Some of Descartes' biographers have written that René's personality was deeply affected by the loss of his mother, and have even speculated that the young boy blamed himself for her death because he did not quite understand—the event having taken place so close to his own birth—that she died sometime after giving birth to her *next* child.

After the death of his wife, Joachim remarried. He took a Breton woman named Anne Morin as his wife, and with her had another son and another daughter (and two other babies who died in infancy). They bought a house in Rennes, where René's older sister joined them in 1610, and where she got married in 1613 to a local man. Until then, René and his older brother and sister were raised by a governess. René Descartes was extremely attached to his governess, who was a devout Catholic. She lived to old age, and Descartes specified in his will that she was to receive a significant amount of money annually for her support.

As a child, René became known as the young philosopher of the family because he had a great curiosity about the world, always wanting to know why things were the way they were. The child grew up in the natural environment of farming and hunting and strolling in the woods. Throughout his life he would make references to the bucolic land of his birth and its natural rhythms. In letters to friends, and in published works, he would describe his childhood memories: the smell of the earth after a rainstorm; the trees at different seasons of the year; the process of fermentation of hay, and the making of new wine; the churning of butter from fresh milk; and the feel of the dust rising up from the earth as it was being plowed. Perhaps it was this early closeness to nature that kindled his interest in physics and mathematics as means for understanding nature and unraveling her secrets.

The Descartes family was wealthy, and René would inherit even more assets directly from his grandfathers on both sides, both of whom had been successful medical doctors. His great-grandfather Jean Ferrand had been the personal physician of Queen Eleanor of Austria, the wife of Francis I of France, in the middle of the sixteenth century. He attained great wealth, which was eventually passed on to his daughter Claude, who married Pierre Descartes. Their son was Joachim Descartes, René's father. In 1566, when Joachim was only three years old, his father, Pierre, died of kidney stones. Pierre's father-in-law, Jean Ferrand, Queen Eleanor's personal physician, performed the autopsy on his son-in-law. In 1570 he wrote up the results of the operation and published them in a scientific paper, in Latin, about lithiasis—the formation of stony concretions in the body. His deep curiosity about nature—even to the point of dissecting the body of his own son-in-law—was passed on to his descendants. While, unlike his grandparents, René Descartes would not pursue medicine as a profession, late in life he would dissect many animals in search of the secret to eternal life.

René Descartes would spend much time during his adult life managing his inheritance. This wealth, including significant land holdings in Poitou, would enable him to pursue his interests without concern about a livelihood. It would afford him to indulge his whim of volunteering for military campaigns as a gentleman soldier without compensation—simply for the thrill of adventure. He would be able to afford luxurious accommodations wherever he traveled, and to employ servants and a valet. His money would even allow him to look after the education of his staff, thus sharing with them some of the privilege of his family's wealth. Descartes would grow up to be a very generous employer and friend.

Despite Adrien Baillet's statement in his very comprehensive 1691 biography of Descartes that the family belonged to the nobility, recent research indicates that, as far as René's life is concerned, this was not true. The renowned French historian Geneviève Rodis-Lewis explains in her 1995 biography, *Descartes,* that the Descartes family gained the rank of knighthood, the lowest status of nobility in France, only in 1668—eighteen years after René's death. According to French law, nobility was conferred on families after three successive generations had served the king in high office. Joachim Descartes certainly held such a position, and some have argued that he had originally sought to become councillor to the Parliament of Brittany in hopes of obtaining nobility status for his descendents. But René Descartes would choose a different direction in life, and so nobility would be conferred on the family only after another member had satisfied the three-generations requirement, years after René's death.

After he remarried, Joachim Descartes spent most of his time in Rennes with his new wife and the children she bore him. He also had interests and family business farther south and west in Nantes, also in Brittany. As they grew older, René and his brother and sister traveled frequently to visit their father. Eventually, René Descartes would see all of western France as his home territory, since time in his later childhood was spent throughout the regions of Poitou, Touraine, and Brittany.

But the frequent travel throughout the region was hard on the boy. In adulthood, Descartes described his health as a child as poor, and recounted in letters to friends that every doctor who had ever seen him as a child had said that he was in such poor health that he would most likely die at a young age. His devoted governess took such good care of him, however, that when he was eleven years old he was healthy enough to be sent away to study at the prestigious Jesuit College of La Flèche.

Chapter 2

Jesuit Mathematics and the Pleasures of the Capital

IN 1603, KING HENRY IV, WHO HAD been raised a Protestant but converted to Catholicism, gave the Jesuits, as a gesture of his goodwill toward this powerful Catholic order, his château and vast grounds in the town of La Flèche, to be used by them as the site of a new college. The Jesuits enlarged the château, and the result was a series of large interconnected Renaissance buildings with spacious, square, symmetric inner courtyards. Entering the grounds, one is struck by the perfect order and symmetry of the checkerboard array of grand courtyards, and by the well-manicured gardens beyond them. This is one of the most impressive college grounds anywhere; today it is the site of a military academy, the Prytanée National Militaire.

La Flèche was well chosen for a college. It lies in Anjou, north of Touraine, in a rich area with forests and gentle rolling hills. The town is attractive, with a river running through its center and lush meadows surrounding it. The gate of the college opens right into the spacious town square. Once students left the grounds of the college, they were in the center of the town with its many places to eat, drink, and find entertainment.

The college was inaugurated by King Henry IV, and opened its gates in 1604. France's brightest students from the best families were encouraged to apply for acceptance. Among the students accepted to the first class to enter this elite new college was Marin Mersenne (1588–1648), who would become Descartes' most devoted and loyal friend.

The college was run as a semimilitary school. The students had to wear a uniform, including breeches with pom-pom fastenings, a fancy blue blouse with large-puffed sleeves, and a felt hat. Each student was given a long list of items he had to bring to school. These included candles, various kinds of pencils, quills, and notebooks, and items of personal use.

According to new research, René Descartes entered the College of La Flèche right after Easter 1607. A college such as La Flèche was an educational institution at which young boys studied before going to a university. As such, it was akin to a modern preparatory school rather than what we call a college today. Descartes' admission to the college was delayed because his health wasn't good enough when he was younger; and until then, he was educated at home by a tutor. René was eleven years old when he started his studies at La Flèche, and he stayed there until his graduation eight years later, in 1615. The tuition at the college was free, but students' families had to pay for room and board and other expenses. At the time Descartes studied there, the college had about fourteen hundred students, who came from all over France. Fortunately, the college was a day's journey from Descartes' home in La Haye, so he could easily visit his grandmother and governess.

Since Descartes was still frail, the family made a special request to the college, asking that its administration take exceptionally good care of his health. The rector, Père Charlet, who was a relative of the Descartes family and whom René would later describe as "almost a father to me," was eager to comply, and gave René unprecedented privileges so that the boy would not feel any stress, in hopes that his health would improve. He thus allowed René to sleep late in the mornings and stay in bed until he felt well enough to join the others in the classroom. This unique arrangement started Descartes on a lifelong routine of waking up late in the mornings and staying in bed, thinking and working, until he was ready to get up and face the day. Most of his life until his last few months in Sweden, with the exception of periods in which he was in an army involved in heavy fighting, Descartes never had a forced wake-up in the morning, but rather arose whenever his body was sufficiently rested.

Life at the college, for everyone else, began at 5:00 A.M. From 5:00 until 5:45, the students said their morning prayers, washed and dressed, and got ready for work. Personal hygiene was very important: students were made to wash well, and if they were sick they were confined to the infirmary. The time from 5:45 to 7:15 was spent on individual work, followed by breakfast. The first classes were held from 7:30 until 10:00. René Descartes was exempt from having to sit through these early classes. At 10:00 A.M., mass was held, followed by dinner, which was the first substantial meal of the day. There followed a period set aside for recreation. Afternoon classes began at 1:30 and lasted until either 4:30 or 5:30, depending on the season. The students spent the remainder of the day—until 9:00 P.M., when they went to bed—having their evening meal and playing games, mostly ball games that were popular in France during this period (including the famous *jeu de paume*). They were also allowed to play cards, a popular pastime, but the involvement of money in any kind of game was strictly prohibited. Just before they retired at 9:00 P.M., they attended a spiritual lecture. On special days, equestrian skills were taught, as well as swordsmanship.

René Descartes joined his fellow students after the morning classes had ended. His special arrangement, allowing him to stay in bed late, made him learn how to study on his own. This was especially useful in mathematics, where he could derive ideas by himself, without having to sit through the material slowly with an entire class. This habit of intellectual independence enabled him to forge ahead swiftly and—a little later in life—to create new mathematical and scientific knowledge on his own without being held back by conventional beliefs.

Students were taught grammar, which meant Latin and Greek. They also studied humanities, rhetoric, and philosophy. Both humanities and rhetoric were centered on the classics. Humanities included the works of the Roman poets Virgil, Horace, and Ovid, while rhetoric was mostly a study of Cicero and the Platonic method of argumentation. The philosophy taught at La Flèche consisted of the works of Aristotle, in the medieval scholastic tradition, as well as logic, physics, and metaphysics.

The Jesuits taught their students the essentials of Greek mathematics: the works of Euclid, Pythagoras, and Archimedes. The Greeks, with their clear, abstract view of nature and its elements, simplified all construction in geometry to figures that could be drawn with only two tools—the simplest tools they could conceive and ones that they believed should suffice to construct any geometrical figure one could imagine or need. These two elementary tools were an unmarked straightedge and a compass.

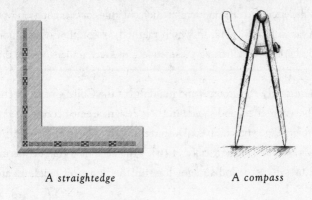

A *straightedge*   A *compass*

The straightedge made angles and straight lines, while the compass was used for making circles and marking off distances. By combining the operations of both instruments, anything that was of importance to Greek mathematicians could be drawn.

Figures drawn with straightedge and compass

Descartes was fascinated by the simplicity of thought and the power of abstraction the ancient Greeks had demonstrated with their geometry, an entire science based on the use of two instruments. The priests who taught this ancient knowledge at the college outlined the entire theory of Euclidean geometry, and explained to the students how to prove a variety of theorems in geometry. Mathematics courses at La

Flèche also included arithmetic, in which the rules of computation were studied, as well as algebra, in which methods of solution of equations were explained—as much as these equations were understood at that time.

Interestingly, the grounds and buildings of the College of La Flèche, perfectly symmetrical and square in their design, seemed constructed as if by straightedge and compass. Descartes' writings later in life—particularly his *Discourse on the Method* of 1637—make it clear that he was impressed by symmetry and straight lines in the design of buildings and towns:

> Thus old cities, which were at their beginnings nothing but small towns, and have become in time large cities, are generally not well designed as by compass [*ordinairement si mal compassées*], at the cost of the regularity of their squares that an engineer traced following his fantasy on the plane. . . . Seeing how the buildings are arranged, here a large one, there a small one, and how they make the streets curved and unequal, one would conclude that it was chance, rather than the will of men using reason, that has thus laid them out.

What the young boy saw on the grounds of his college and learned from his study of Greek geometry using straightedge and compass became important elements in shaping his philosophy and his mathematics. With enough imagination, one can trace the origins of the Cartesian coordinate system to the perfectly symmetric design of the grounds of Descartes' College of La Flèche.

During his time at La Flèche, Descartes and the other students at the college took part in a most unusual ceremony. King Henry IV was considered the founder of the college, and the gate to the college bore (and still does today) the letter *H* on each side, in honor of the king, as did most buildings on the campus. In his will, the king had specified that his heart, that of his queen, and those of their heirs were to be buried at the Church of Saint Thomas on the grounds of the college, his former Château of La Flèche.

Henry IV, whom history would remember as one of France's most benevolent monarchs (he's the king who promised a "chicken in every pot" in his realm for Sunday meals), had difficulties throughout his rule with tensions between Protestants and Catholics, the latter never quite trusting him completely even though he was nominally a Catholic. Things became more dangerous for the king when in 1610 he allied himself with several German Protestant princes against Catholic Spain. On May 14 of that year, the king's carriage was passing through a particularly busy street in Paris, the rue de la Ferronerie. As traffic on the street momentarily came to a halt, a Catholic fanatic named Ravaillac rushed onto the royal carriage and stabbed the king in the chest. He died almost instantly.

After the king's assassination, the ritual prescribed in the king's will began. It started with prayers for the soul of the dead king held on May 15 at the church of the College of La Flèche. The king's body was embalmed in the Louvre palace, and the heart was carefully removed. The heart was kept for three days in the Jesuit chapel in Paris, and then transferred to the care of Père Provincial Armand, who gathered a group of twenty leading Jesuits and a larger number of the king's knights to accompany the heart to its final resting place in La Flèche.

The college chose a group of twenty-four of its most distinguished students to join the procession as the party carrying the king's heart arrived in town. Among these twenty-four students was René Descartes. The procession came to order at the central square in town, and now included also archers and royal guards, who had joined it. From the town of La Flèche, the entire assemblage marched to an open field and stopped. Torches were lit, and in a solemn ceremony, the heart was transferred from Père Armand to the duke of Montbazon. The procession then continued on to college grounds and into the Church of Saint

Thomas. Inside the church, the duke raised the heart for all to see, and in a communal prayer, it was placed in an urn forever to remain in the church. This ceremony was deeply etched into René Descartes' memory.

After his graduation in 1615, Descartes moved to Poitiers to study law at the university. He spent an uneventful year there, since he was not interested in the law. Much of his spare time was devoted to practicing his swordsmanship skills learned at La Flèche. He received his doctor of law degree in 1616, and after spending a pleasant summer with his family in Rennes, riding horses and enjoying the outdoors, he moved to Paris. (In 1985, Descartes' thesis for his law degree was discovered at the University of Poitiers. The date of acceptance of the thesis was November 10, 1616. When they found the thesis, scholars were struck by its date. This day—November 10, and sometimes the night of November 10–11, or November 11—reappears as if by magic as the date of many key events in the life of Descartes.)

Descartes' family did not take well to his expressed wish to move to Paris after his graduation. Even though René's health was now much better than it had been in his childhood, the family was concerned about the prospect of his living alone so far away. Joachim Descartes was especially worried about his son moving to the big city, but René assured him that this was what he wanted to do and that it was an important step toward his future. At twenty, he argued, he was old enough to live on his own and find his way in the world. The father finally agreed to let him go—but only on the condition that he move to Paris with servants and a valet to assist him. Throughout his life, Descartes would never be without a valet. He chose them well—his valets were always extremely loyal to him and would literally risk their lives for his safety and well-being.

The city of Paris has always been the center of life for the French, who have been flocking there for centuries from small towns and villages throughout the land. They come to seek a better life, economic rewards, and culture. But many come to Paris for the sheer excitement it offers. In the days of Descartes, this impulse was no different than it is today.

Descartes lived during the time of d'Artagnan and the swash-buckling adventures depicted in Alexandre Dumas' *Three Musketeers*. Looking at seventeenth-century French paintings and engravings gives us a good idea about the styles of the time of Descartes: richly colored flowing silk garments with voluptuous folds, plumed velvet hats, shoes with ornate silver buckles. Descartes loved to dress in style, and he carried his polished sword with him wherever he went, as did many young gentlemen of his time. He was intent on learning as much as he could from "the great book of the world," as he called it. He was on a search for truth about life and the human experience. And what better place to explore life than the city of Paris?

By the time he finally moved to Paris, Descartes' health was excellent. He had lost the paleness of his youth, and he no longer suffered from whatever ailments had plagued him in childhood. Feeling well for the first time in his life, and finally having finished his education, he was eager to experience life. In Paris he met up with many of his old friends from La Flèche, who had congregated in the capital, and he also made some new ones.

Descartes' early days in Paris were a carnival of drinking, gambling, and merrymaking. He learned very quickly that he had a knack for card games and other games of chance—now, unlike at college, played for money. He was very successful at these games, winning significant amounts of money, which made him even more popular with his old and new friends alike. Descartes was constantly surrounded by people, and life for him was a continuous, endless party. These young men paraded down the streets of Paris, pursuing beautiful women.

When he was a child, René had a crush on a young girl who had a lazy-eye problem. This childhood memory made him always focus his attention on women's eyes. As a young man in Paris, he found himself especially attracted to women who had beautiful eyes.

Among René's new friends in Paris was Claude Mydorge, who had been the treasurer of the city of Amiens, and who enjoyed a reputation as a first-rate mathematician. Mydorge was a dozen years older than Descartes, a man of the world, and he had an especially engaging and vivacious personality and a good sense of humor. Descartes was attached to him, and the two men spent many hours together enjoying themselves as well as discussing mathematics. Among René's acquaintances from La Flèche was Marin Mersenne. Mersenne had by then finished a course of study at the Sorbonne and had received the habit of the Minim order on July 17, 1611, in the monastery of Nigeon, near Paris. The Order of the Minim Brothers was founded in 1435 by Saint Francis of Paola in Calabria, Italy. To the Minims, humility is the primary virtue, and the name Minim derives from *minimi*, since they view themselves as "the least" of all the religious.

Six months after Descartes arrived in Paris, Mersenne was ordained a priest and became a friar of the Minim monastery at the Place Royale (today's Place des Vosges). Descartes would often visit Mersenne, who had a very wide range of scientific and mathematical interests, and talk with him and explore new ideas. Mersenne quickly became Descartes' closest friend. According to Baillet, the stimulating interactions Descartes enjoyed with Mersenne served as a counterweight to the general lack of purpose that characterized his early days in Paris.

Descartes also loved music. Some scholars have drawn a connection between his interest in music and his great ability in mathematics. At any rate, Descartes and his many friends pursued musical interests together, attending concerts and performances throughout the capital.

After about a year of play and enjoyments of all kinds, René Descartes felt the need to become more serious. In Paris he had come in contact with new ideas. French intellectuals were studying Greek geometry, trying to embellish the works of the ancient Greeks. Euclid's *Elements* had by then been extended to include three more volumes beyond the original thirteen. Physics, too, was developing, as scientists were studying the nature of falling objects and exploring the riddle of gravity. Descartes longed to study these new ideas, and his feverish social life became an impediment to achieving this goal. To concentrate on his work, the young man now felt he had to distance himself from his many friends—but they made it hard for him to do so. Whenever Descartes stayed home to read and work, his friends would come over to his apartment and beseech him to join them on the streets and at the nightspots of the city.

In desperation, Descartes took a drastic measure—he secretly moved and did not give his friends his new address. He needed to be in an area where people would not recognize him whenever he went out for a walk, or recognize his servants or valet when they shopped or ran errands for him. So Descartes moved outside the walls of the city, to the neighborhood around the ancient Church of Saint-Germain-des-Prés. He loved this part of Paris because it was peaceful and quiet, and more rural. There were open fields here, to which young men would occasionally come to duel, an activity discouraged by the authorities. Descartes began to view himself as a spectator rather than an actor in the drama of daily life around him. For over a year, no one saw Descartes. His friends worried, and they suspected that he might have left Paris to return to his father's house in Brittany. They complained to one another about his incivility in leaving without saying good-bye. Some of them made inquiries in Rennes, but learned that he was not in Brittany, nor in Touraine or Poitou. They kept looking for him in the capital, but found him at no ball or banquet or reception. His friends were close to giving him up as lost.

Hiding out in a new part of the city worked for Descartes for some time. But his friends were still searching for him everywhere. One day, Descartes' valet was recognized on the street by one of these acquaintances, who then secretly followed the valet out of the city walls and into the area of Saint-Germain. The man waited for the valet to disappear up a flight of stairs, and then followed up the same stairway. He looked through the keyhole into Descartes' bedroom.

The man observed Descartes lying in bed reading, then sitting up in bed to write in a notebook, then lying down again to read, and a little later sitting up again to write in the notebook. The friend realized that Descartes was immersed in work he seemed determined to keep from the rest of the world. He would not disturb him, now that he understood Descartes' need to withdraw from the excited life of the capital to pursue his private work, and he quietly left.

Scholars believe that Descartes likely wrote his enigmatic *Preambles*, which included the statement "I advance masked," in his hideaway in Saint-Germain. Descartes' *Preambles* continued:

> The sciences are now masked; the masks lifted, they appear in
> all their beauty. To someone who can see the entire chain of
> the sciences, it would seem no harder to discern them than to
> do so with the sequence of all the numbers. Strict limits are
> prescribed for all spirits, and these limits may not be trespassed.

If some, by a flaw of spirit, are unable to follow the principles of invention, they may at least appreciate the real value of the sciences, and this should suffice to bring them true judgment on the evaluation of all things.

Descartes' ideas about the "chain of the sciences" and the sequence of the natural numbers bore a striking resemblance to mysterious writings of a mystical nature that had begun to appear in Europe in the early part of that same decade. The authors of these treatises on science and mathematics had remained anonymous.

Two years after he arrived in the French capital, René Descartes was ready to move on. He had always enjoyed a fast lifestyle, and he liked to fence and ride horses; now he longed for a life of action. Descartes had heard that up in Holland, Maurice of Nassau, the new prince of Orange and the Protestant champion in the religious wars, was gathering men from several countries—including two French regiments—and training them in his camps for war with the gathering Catholic forces of Spain and Austria.

Although he was a Catholic, Descartes was interested in joining Prince Maurice's army. He felt that he could learn much about the art of war from the prince and his generals, and religion did not weigh in his decision, perhaps because he would join as a volunteer and would not have to fight if he chose not to do so. Descartes sent his servants back to his father in Rennes and, taking only his valet with him, traveled to Breda in southern Holland to volunteer his services. He would learn the art of war, but would not be paid for his military service—other than a token single gold doubloon. Unpaid, he would retain many freedoms, and these rights would allow him to pursue his research in mathematics and science, and to try to uncover their hidden meanings. Descartes would use the army as a vehicle for travel and adventure. In the words of Baillet, Descartes would use the army as "a passport to the world."

Chapter 3

The Dutch Puzzle

"QUID HOC SIGNIFICAT?" THE YOUNG French soldier asked the slightly older Dutchman standing next to him, addressing him in Latin, the lingua franca of the educated throughout Europe. They were both among a group that had crowded around a curious poster attached to a tree trunk in the main square of the city of Breda, on the morning of November 10, 1618. "What does this mean?"

The Dutchman was from Middleburg and had just finished his studies in medicine and mathematics, and was hoping to take up the position of assistant principal of the Latin School of Utrecht. He had come to Breda to help his uncle slaughter his pigs, and was also hoping to find a wife. He took a long look at the young French soldier facing him. René Descartes wore the distinctive uniform of a volunteer in the army of Maurice of Nassau, the prince of Orange.

The Dutchman also noticed that the soldier was wearing a fancy plumed green hat, and that a silver sword hung from his hip—not the usual musket other soldiers carried. He looked not much older than twenty-two or twenty-three, was of medium height, slightly less perhaps, with longish thick and wavy dark brown hair, and a mustache and goatee to match. And he had piercing, earnest brown eyes.

The soldier was looking at him expectantly. "It is a mathematical puzzle," the Dutchman answered.

"I can see that," responded Descartes, "but what exactly does it say? I do not understand Flemish."

The man took out a piece of paper and a pencil and began to copy the geometrical designs on the poster, labeling each in Latin, rather than Flemish, and then translating into Latin the paragraph written below the diagram. He handed the paper to the young soldier and said, "They want you to prove this statement," pointing to the last sentence. Descartes looked intently at the paper in his hand, and the man added: "And I suppose you will give me the solution, once you have solved this problem?"

Descartes turned quickly from the paper and gazed intently at the man.

"Yes, of course I will give you the solution," he said with determination. "Would you please give me your address?"

The Dutchman offered him his hand and said, "Isaac Beeckman is my name."

"René Descartes," said the soldier. "Or René *le Poitevin*, as they call me, since my family comes from the French region of Poitou; although I was actually born in Touraine."

They shook hands, and Beeckman told Descartes that he was from Middleburg but was staying in Breda to help his uncle with his pigs. He gave him his uncle's address and they bade each other good-bye. From his journal, which was discovered in a Dutch library in 1905, and from other sources, we know that Beeckman did not believe that the young soldier would be able to solve the puzzle.

The next morning, as Isaac Beeckman was about to have breakfast at his uncle's house, there was a persistent loud knock on the door. The servant opened the door and let in the young soldier. He was accompanied by his valet. Descartes showed Beeckman his solution to the Dutch puzzle. Beeckman, who was a competent mathematician, was amazed by the soldier's solution to a very difficult mathematical problem. He had not expected a random person to solve a problem that many trained mathematicians and professors could not. Descartes' brilliant solution cemented a friendship between the two men. It was also a watershed event in the life of the young Descartes, since this was the moment he first realized that he was a very gifted mathematician.

The puzzle Descartes solved and demonstrated to his new friend was not an isolated problem suddenly appearing on a poster in southern Holland. The seventeenth century saw a revival of the classical geometry of ancient Greece as educated people throughout Europe sought intellectual challenges and the hidden meanings of mathematics. Ancient Greek texts were being republished in Latin, foremost among them Euclid's classic volumes of the *Elements*, written in Alexandria about 300 B.C. Euclid's work was, in fact, the most important textbook published on the new printing presses invented less than a century and a half before Descartes' time. Other ancient texts, such as Diophantus's *Arithmetica*, written around A.D. 250, were also being printed in seventeenth-century Europe. It was on the margin of a copy of this book that Pierre de Fermat (1601–65) wrote his famous Last Theorem, which would haunt mathematicians and amateurs alike until its final, dramatic proof late in the twentieth century.

These newly republished mathematical texts enabled a revival of the study of geometry in the schools and universities of Europe, and with these new publications arose a whole class of intellectuals who avidly pursued new solutions to ancient problems, challenging one another to solve problems they proposed and publicized on posters placed in public places. The problem Descartes solved on November 10, 1618, was one such example of a challenge issued by a mathematician through a public posting. Similar challenges issued a century earlier by mathematicians living in northern Italy had led to great developments in the area of algebra, resulting in the solution of complicated equations that the ancient Greeks and the medieval Arab mathematicians who followed them had not been able to achieve.

We don't know exactly what the problem was that Descartes solved and showed Beeckman in November 1618. We do know that it involved angles in a geometrical drawing, and that it was a very difficult problem. But one unusual, and perhaps incidental, aspect of Descartes' geometrical thought apparently impressed Beeckman very much. Descartes may have raised this particular issue on meeting Beeckman, even before he presented him with his solution to the problem from the poster the next day. In his journal, Beeckman described it on the day after he met Descartes:

> *Angulum nullum esse male probavit Des Cartes:*
>
> Yesterday, which was November 10, 1618, at Breda, a Frenchman from Poitou tried to prove the following: "In truth, there is no such thing as an angle." This was his argument: "An angle is the meeting point of two lines at one point, so that the line *ab* and the line *cb* meet at point *b*. But if you intersect an-

gle *abc* by the line *de*, you divide point *b* into two parts, so that half of it is added to line *ab*, the other half to line *bc*. But this

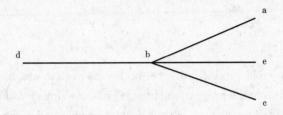

contradicts the definition of a point, since it has no size. Hence, there is no such thing as an angle.

This, of course, is unlikely to have been the complicated mathematical puzzle Descartes solved, but rather Descartes being clever with Beeckman, showing off his mastery of Greek geometry with its axiomatic definitions—to the point of making an absurd argument. The *point* in the angle does not, in fact, get cut into two halves—precisely because it has no size. Descartes was philosophizing, and Beeckman was impressed with his mastery of the intricacies of Greek geometry.

At the time Descartes showed Beeckman his solution to the Greek geometrical problem, the two areas, geometry and algebra, were considered two different parts of a wider, and somewhat nebulous, field called *mathematics*. Geometry was about straight lines and triangles and circles—idealized visual images of the elements of the physical world. Algebra, however, was the study of *equations*—symbols and numbers on two sides of an equal sign, which had to be solved to result in some meaningful quantity. No one had imagined that these two fields could be unified into a larger discipline. But within two decades, René Descartes would do just that.

Descartes told Beeckman that he was hoping to go to war. Beeckman worried about this prospect, and hoped that his new friend would stay in the area and that the two of them could meet often and work together on problems of mathematics and science. After he left Isaac Beeckman on November 11, 1618, and returned to his army camp, Descartes realized that he was not yet going to be sent to war. He stayed on in Prince Maurice's army for several months while the troops remained stationed outside Breda. Since he was, after all, a volunteer, Descartes did more or less what he wanted to do. He spent time learning Flemish, so that he would never again have to ask strangers for translations. Descartes had excellent facility with language—he had perfect mastery of French and Latin—and before long, he could understand Flemish well and even speak it with some fluency. This new language ability also gave him fluency in other, related, Germanic dialects. Descartes was proud of his new achievement, and on January 24, 1619, he wrote to Beeckman from his army camp: "I devote my time here to painting, military architecture, and the study of Flemish. You will soon see the progress I have made in this language, when I come to see you in Middleburg, God permit it, at the beginning of Lent." Descartes could not have known it at the time, but his new mastery of Flemish and related Germanic dialects would literally save his life.

Descartes' solution of the Dutch puzzle excited him about mathematics. For it demonstrated to him that he had a unique gift. He began to believe that mathematics held the secret to understanding the universe. He stayed most mornings in bed at camp reading and writing about mathematics and exploring its applications. He worked out ancient Greek problems in geometry, but he soon concluded that the power of geometry transcended pure mathematics: geometry held the secret to all creation.

Three years before they met in Breda, Isaac Beeckman had penned an article—which was discovered in his journal—about the mathematics of music. In this article, Beeckman tried to use Greek geometry to explain the harmonics of a vibrating string. While Beeckman's analysis was not very deep, Descartes did not show any lack of respect for his new friend's work. The two men worked together, trying to create a theory of music based on mathematics. They also worked on mechanics and on pure geometry. Beeckman would suggest problems, and Descartes would solve them using his brilliant mathematical abilities. By now Beeckman was back home in Middleburg, and Descartes would visit him whenever he could. When they were not together, the two friends exchanged ideas through letters.

On March 26, 1619, Descartes wrote his friend from his camp in Breda. He unfolded his plan to invent a method of solving a very wide variety of problems. He wrote: "I desire to give the public not an *Ars brevis* of Lull, but a science based on new foundations." Descartes alluded here to the work of Ramon Lull (c. 1235–1315), a medieval mystic who was born on the Spanish island of Majorca and wrote 260 books, among them the *Ars brevis* ("Brief Art") Descartes mentioned. Lull's works were a mixture of Cabbala and mysticism whose elements were ways of combining letters and numbers in an attempt to extract new knowledge about the universe.

A month later, on April 29, 1619, Descartes again wrote to Beeckman about the work of the Majorcan mystic: "Three days ago I met at an inn in Dordrecht a learned man and discussed with him the

Ars parva [Ars brevis] of Lull. He said that he could use the *Art* so successfully that he could discuss any topic whatsoever for an hour; and if one then asked him to speak about the same topic for another hour, he could do so without repeating what he had already said, and so on for another twenty hours."

Descartes was interested in the supposed power of Lull's mystical methods of obtaining knowledge. He asked Beeckman about it, and the latter, who had read some of the works of Lull, explained to him that Lull had invented a wheel on which the nine letters *B, C, D, E, F, G, H, J,* and *K* were arranged. These nine letters stood for attributes of creation (which were akin to the ten attributes of God, the *Sefirot,* in the Jewish Kabbalah). By producing permutations of these letters using circles revolving within other circles in a geometrical device, new concepts could be derived. Descartes' letters to Beeckman in 1619 give us the first indication of the young man's curiosity about mystical methods and ideas.

Lull's medieval magic would be reflected in the teachings of a secret society emerging in the early seventeenth century, and Descartes would find himself in the midst of these explorations of science and mysticism.

When Descartes returned to his camp in Breda, he began to suspect that he might never have a chance to see the military action he longed for. The prince of Orange had signed a truce with his enemies, and as part of this arrangement, had made a commitment not to engage in battle for a period of twelve years.

Descartes felt betrayed by this truce since he felt that the promise he had been given when he volunteered to serve the prince—that he would see military action—would likely never be met. Descartes had known since the previous year, 1618, that important political events were taking place in Bohemia and in Germany, and that these events could lead to war. By that time, the winds of religious war had been blowing in Europe for an entire century.

When Martin Luther (1483–1546) posted his ninety-five theses on the church door in Wittenberg in 1517, launching the Lutheran branch of Christianity and hence Protestantism, religious confrontations erupted all over Europe. Luther's act is generally taken as the starting point of the Reformation, a movement for reform of the doctrines and practices of the Roman Catholic Church. The controversy between Catholics and Protestants had a strong political aspect as well, since religion and nationality tended to correlate across the Continent. By 1530, the rulers of the German states of Saxony, Hesse, Brandenburg, and Brunswick had been won over to the reformed belief, as were the kings of Sweden and Denmark. Consequently, these rulers broke away from Catholicism and made the churches in their realms conform to Protestant principles.

Calvinism, a form of Protestantism based on the teachings of John Calvin (1509–64), which includes the idea of predestination, was founded in 1536 and took hold in parts of France, western Germany, Holland, Switzerland, and Scotland. Since the rest of Europe remained Catholic and loyal to the pope, the continent was deeply divided. In France, the Wars of Religion took place in the mid-to-late sixteenth century between Catholics and the Huguenots—French Protestants. These wars were complicated by the interventions of Spain, Savoy, and Rome on the side of the Catholics, and of England, Holland, and several German principalities on the side of the Huguenots. By the time Descartes was growing up, however, there was relative peace. But this peace was not to last for long.

In 1613, a young German prince, Frederick of the Palatinate—a region in southern Germany that included part of the Rhine and the city of Heidelberg—arrived in London to marry Elizabeth Stuart, the daughter of King James I of England and Anne of Denmark, and granddaughter of Mary, Queen of Scots. Many in Europe saw in this royal wedding the forging of an important alliance between England and Protestant forces on the continent. But in fact, King James I had every intention of remaining neutral in the religious conflict that was brewing on the continent, and was hoping to balance the apparent favoring of Germany that might have been implied by his daughter's wedding by forging a similar union with Catholic Spain.

Frederick took his bride to Heidelberg, sailing down the Rhine, and they established their home in Heidelberg castle, where they lived happily and in peace for five years. But then things began to change. A strong movement on the continent saw in Frederick a potential leader who could unite the Protestant forces in Europe against the Catholic forces of the Austrian Habsburg empire, whose capital was Prague.

The Habsburg emperor Rudolf II, a benevolent and tolerant ruler who had moved the center of the Habsburg empire from Vienna to Prague, died in 1612, precipitating a power struggle for his succession. Prague under Rudolf had been a flourishing city in which new ideas were developed and learning was promoted. Rudolf had a strong interest in magic and mysticism, and under his rule, Prague became a center for alchemical, astrological, and magico-scientific studies of all kinds. Jews practiced Kabbalah in this city, and were treated equally and without discrimination. Many others studied the occult and the magical. Rudolf's palace had "wonder rooms" in which mechanical magical arts were practiced, including inanimate objects that were made to seem to talk, numerological analyses, astrological predictions, and alchemical wonders. In the 1580s, the famous English mystic and mathematician John Dee (1527–1608) spent several years in Prague spreading his magical knowledge. Dee's ideas would form the foundation of knowledge pursued by a mystical secret society. This new society would also develop a strong political agenda for reform in Europe, and its members would see their great hope for political salvation in the person of Prince Frederick of the Palatinate.

Descartes, a devout Catholic, was serving in the Protestant army of Maurice of Nassau, who happened to be Frederick's uncle. Descartes followed closely the events in Bohemia that resulted from the power vacuum following the death of Rudolf II. He knew that on May 23, 1618, the defenestrations had taken place in Prague, in which the rebellious Bohemians threw their Habsburg governors out of the windows of their castle. The state of Bohemia then raised two armies, one headed by the count of Thurn and the other by the count of Mansfeld. Duke Maximilian of Bavaria, a Catholic, reacted by raising an army to help the Austrians of the Holy Roman Empire against the rebelling Protestant Bohemians. He allowed eight thousand men and two thousand horses to pass through Bavaria on their way from the Catholic Spanish Low Countries (allied with the Habsburgs) to prepare to attack Prague. Descartes, eagerly following these developments, could not hold back his anticipation of battle.

Since it was clear that Prince Maurice—because of the treaty he had signed—could not take his troops there, Descartes made the decision to quit his commission and travel to Germany on his own, with the intention of joining someone else's army. He was hungry for travel and adventure, and he hoped to see action.

But Descartes did not want to leave Breda without saying good-bye to his friend Isaac Beeckman. On April 20, 1619, Descartes wrote to Beeckman: "I ask of you at least to respond to me by an intermediary, since he is at my service, and to tell me how you are doing, what occupies your time, and are you still concerned with getting married?" Beeckman responded within a day. He was working on mathematics—on the ideas Descartes had discussed with him at their last meeting.

And, no, he hadn't found a wife yet. It was unusual for the two of them not to see each other. When Descartes was in Breda—from the time they first met, in November 1618, until January 1619—the two friends met almost every day. But Descartes was now with the troops and Beeckman was in Middleburg, so a letter had to suffice.

Descartes wrote to Beeckman again on April 23, 1619, saying that he wanted to see him. He told him that his soul was already traveling. Borrowing Virgil's words in the *Aeneid* (III.7), Descartes wrote: "I don't know yet where Destiny will guide me, nor where I shall stop along the way." He continued:

> Since the threat of war no longer leads with certainty to Germany, and I fear that I would find there many armed men but no combat, I will stroll, while waiting, through Denmark, Poland, and Hungary, until I find in Germany a road clear of brigands and highwaymen to take me surely to war. If I stop somewhere along the way, as I hope to do, I promise you to put in order my work on *Mechanics* or *Geometry*, and to honor you as the promoter of this work.

After saying good-bye to Beeckman, on April 24, 1619, Descartes traveled north. He left Amsterdam on April 29 and headed for Copenhagen, where he stayed for some time, and also visited the rest of Denmark. From there he continued east to Danzig. Some weeks later, Descartes headed south, traveling through much of Poland, and entered Hungary. Then he turned west and reached Frankfurt on July 20, 1619. He arrived there to witness an important event: the election of the new Holy Roman Emperor. The king of Bohemia, Ferdinand II, was elected emperor on August 28 and crowned two days later. Descartes was present at this magnificent ceremony. The new Holy Roman Emperor was given the globe and scepter of Charlemagne, and was passed Charlemagne's sword, which he raised high for all to see.

Bohemia no longer had a king, and the rebellious Protestant Bohemians—in clear defiance of their Austrian overlords—decided to *elect* a king, rather than submit to Habsburg rule. By that time, all of Protestant Europe had been looking up to Frederick, Elector Palatine of the Rhine (so called because he was one of the German princes with the privilege of electing the Holy Roman Emperor), as their potential leader. The Bohemians easily elected Frederick of the Palatinate to become Frederick V, king of Bohemia, and offered him the crown.

The prince of Orange, Frederick's maternal uncle, supported his assuming the role for which he had been elected; but James I of England, Frederick's father-in-law, who was very worried about this development, told Frederick that he thought that he was too young and inexperienced to assume what he considered the precarious role of king of a nation about to face a massive military attack. However, James's daughter, Frederick's wife, was eager to become a queen, and she pressed her husband to take the throne. Frederick should by all rights have listened to the advice of his father-in-law, for in order to remain king he would badly need England's support, and James I was clearly uninterested in helping him. As it turned out, Frederick listened to his wife and uncle and allowed himself to be crowned. Three days after the coronation of her husband, Princess Elizabeth of England was anointed Queen Elizabeth of Bohemia.

Predictably, the Austrians were outraged by the coronation of the new king and queen of Bohemia, viewing it as a rebellion by a province. Duke Maximilian of Bavaria then resolved to go to war in order to help the Austrians overthrow the new king. "During all these movements, Descartes enjoyed the tranquillity that came from his complete indifference to all these foreign affairs," Baillet tells us. Descartes stayed on in the region to work on mathematics, but soon decided to pass through

Bohemia and observe the imperial and Bohemian armies fighting each other. He saw cities and towns pass from one hand to another, and then back, as one army or the other enjoyed success in battle.

Descartes then made his decision: he would join the army of Maximilian, the duke of Bavaria. He would again be a volunteer and would not carry a musket, only his sword. Descartes would enjoy all the privileges he had enjoyed in Prince Maurice's army, including retaining his valet and having as much free time as was possible. With these terms signed, Descartes headed south, and in October 1619 reached the small town of Neuburg an der Donau, perched on the banks of the Danube in southern Germany about halfway between Munich and Nuremberg. This was where Duke Maximilian's forces were encamped for the winter. The clouds gathering over Europe that fall of 1619 would signal the beginning of the Thirty Years War. It would end only in 1648, with the signing of the Peace of Westphalia.

Chapter 4

Three Dreams in an Oven by the Danube

Now settled with the troops on the banks of the Danube, Descartes prepared to spend the winter working on science and geometry. In his *Discourse on the Method*, published in 1637—some years before the Peace of Westphalia put an end to the Thirty Years War—Descartes wrote:

> I was at that time in Germany, where the occasion of the wars that have not yet ended called me. As I left the coronation of the emperor to return to the army, the onset of winter forced me to remain in quarters that offered no opportunities for conversation or entertainment, and, by good fortune, also no worries or passions to trouble me. I remained, each entire day, alone in a closed

oven [*poêle*], where I had all the time, at my leisure, to entertain myself with my own thoughts.

The "oven" Descartes mentions was a room heated by a large central woodstove. The stove was used both for cooking and for heating in
winter. Descartes' thoughts in the oven took place in November 1619.
And we know that, in particular, something happened to Descartes on
the night of November 10–11, 1619, while he slept in his oven—something that would bring about a transformation of his life and a sharpen-
ing of his thinking.

Baillet, who had access to the material in the inventory made after
the death of Descartes, and who, in particular, read Descartes' manuscript described in the inventory as "Item C1: A small parchment register entitled *Olympica*" (now lost, but also copied by Leibniz), detailing
the events of that fateful night, tells us the story.

The date of the event is significant: November 10 was the first anniversary of Descartes' initial meeting with Beeckman, which resulted
in his solving the geometrical problem from the poster in Breda and
sparked his love of mathematics. It was also the third anniversary of the
acceptance of Descartes' law thesis at the University of Poitiers. But a
year after meeting Beeckman and making the realization that he was a
gifted mathematician, now twenty-three years old, Descartes still had
not decided which course his life should take.

In the *Olympica*, Descartes described the day: "10 November 1619,
I was filled with enthusiasm since I was on the verge of discovering the
foundations of an admirable science." According to Baillet, this is what
happened during the night. November 10–11 was Saint Martin's Eve,
which was customarily a night of drinking and debauchery. The other
soldiers all went out drinking in celebration of the saint's feast. But
Descartes said much later that he consumed no alcohol that evening,
and in fact had not drunk any wine for three months before this day. So

we are left deprived, perhaps, of the simplest explanation for the unusual nature of what Descartes experienced next.

He went to bed in his oven and had a sequence of three very vivid and very powerful dreams. Arguably, Descartes experienced in the night of November 10–11, 1619, the three most famous and most frequently analyzed dreams in history. In fact, his dreams that night would change history, for they would eventually lead to the first unification of two major branches of science—the wedding of geometry with algebra—which would also give us the Cartesian coordinate system, the basis for so many modern technologies.

Descartes did not say when he went to bed. But as soon as he fell asleep, he had the first dream. In this dream, Descartes was walking the streets and was beset by a violent wind raging through a town, bending the trees and howling through doorways. The wind was so strong that he had to lean and walk hunched to the ground. He felt the pain from this great natural violence and was desperate to find shelter from the storm. Suddenly he saw a college, his own College of La Flèche, and on its grounds, the church he knew. He wanted to enter the church to pray, but remembered that he had just passed a person without salutation and wanted to retrace his steps to excuse himself. But the violent wind pushed him strongly "against the church." At this moment he saw in the courtyard of the college outside the church another person he knew, and that person called to him by his name. He spoke to him politely, asking him if he would like to go to see Monsieur N., since that man had a melon to give him, a melon that had been brought from a foreign land.

What struck Descartes most, as he described it in the *Olympica* as reported by Baillet, was that he noticed that everyone around him was suddenly walking very straight, while he was still curved down to the ground and unsteady on his feet because of the wind. Suddenly the wind diminished significantly, Descartes straightened up, and woke up from his dream. Waking up, he felt "a deep sorrow that made him believe it was the work of an evil spirit that wanted to seduce him." He prayed to God for protection from the unknown forces he feared were bent on punishing him for his sins, for he felt that his offenses must have been very serious for him to have encountered the wrath of a storm from heaven upon his head. He spent the next two hours awake, "thinking about the good and the bad in this world."

Descartes had been sleeping on his left side. He changed sides and fell asleep again. Then he experienced his second dream. In this dream, Descartes was in a room, and he felt the room fade, and suddenly he heard a terrible sharp bang, which he believed to be thunder. The storm of the previous dream had returned, but it felt like a hallucination. The tempest could not reach him—he was protected in the safety of his room. Then Descartes saw the room fill with magnificent sparkles of light, and he woke up again.

In the third dream, Descartes was sitting at his desk with an encyclopedia (or a dictionary, according to one interpretation) in front of him. When Descartes put forward his hand to reach for the encyclopedia, he found another book, entitled, in Latin, the *Corpus poetarum*. He opened this book to a random page and found there a poem, Idyll XV, by the Roman poet Ausonius. He began to read its first line: *"Quod vitae sectabor iter?"*—What road shall I follow in this life? Then an unknown person appeared and presented Descartes with another Ausonius verse, titled *"Est et Non."* But as soon as Descartes tried to take hold of the *Corpus poetarum*, it disappeared. Instead of it, he found again his encyclopedia—but this time it was not quite as complete as it had been earlier. Then both the unknown person and the book disappeared. Descartes continued to sleep, but he was now in a heightened state of consciousness that told him that what he had just experienced was a dream. He was able to interpret the dream while still asleep.

Descartes understood the encyclopedia to represent all the sciences put together. The *Corpus poetarum* he took to stand for "philosophy and knowledge joined together." The reason for that assumption was Descartes' belief that poets—even ones writing "silly" verses—all had something to say that was no less valuable than the works of philosophers. The poets brought Descartes his revelation, but also his "enthusiasm" experienced the day before the dreams—a euphoric feeling of discovery. Descartes took the poem *"Est et Non"* to represent "the Yes and the No of Pythagoras, understood to mean *truth* and *falsehood* in the secular sciences."

Descartes interpreted the melon in the first dream as the charm of solitude. He interpreted the violent wind of his first dream, pushing him hard against the church, as the evil spirit, bent on forcing him into a place to which he was going to go of his own free will. It was for that reason that God did not allow him to advance too far away from his destiny, even though the evil spirit was directing him into a holy place.

Descartes interpreted the great sound of the second dream, a lightning and thunder that turned into sparkles of light inside his room, to represent the spirit of truth that came to possess him. Descartes now had an answer to his question posed in the third dream by the first poem of Ausonius: Which road should I choose in life? And the answer was that his mission in life was to unify the sciences. Having had his first revelation, upon solving the Dutch puzzle, that mathematics was his gift, Descartes now understood that unifying the sciences meant work in mathematics. His philosophy—his search for absolute truth and his principle of doubt—which he would develop in the years to come, was his attempt to impose reason and rationality on the universe using the principles of logic and mathematics. His philosophy would thus be inexorably linked with his geometry. But first, then, Descartes' charge was to develop his geometry—to bring its ancient Greek principles to the seventeenth century, in which he lived, and ultimately to bequeath to the world the new science he would create: *analytic geometry.*

Descartes spent the entire following day reflecting on his dreams. He thought that the spirit in the dream, a "genie" as he called it, had inserted the dreams into his head even before he had gone to bed, and that human elements had no effect on anything that followed. His dreams had been completely predetermined. Descartes reflected long about the dreams, and asked God to make him know his will and to conduct him toward the truth. He vowed, in return, to make a trip to Italy for a pilgrimage to one of the most important religious sites in the land: the shrine of Loreto, believed to include the cottage from Nazareth in which the Holy Family had lived. He wanted to leave for Italy by the end of November, but in the end, he would travel there only four years later. Descartes remained in the solitude of his oven and made a vow to write a treatise and finish it by Easter 1620. According to Baillet, the *Olympica* was written at this time, but the biographer felt that there was little order in the mysterious fragments of this manuscript, so that Descartes' vow to write a text must have meant a more significant manuscript than *Olympica*, and that this larger work was merely announced in the *Preambles* and the *Olympica*.

On November 11, 1619, Descartes described in his parchment register the dreams he experienced the night before, and his interpretation of these dreams. In their essence, the copy made by Leibniz (now held with the rest of his papers in the archives of the Gottfried Wilhelm Leibniz Library in Hanover) and Baillet's account agree with each other. The Leibniz copy of *Olympica* was discovered in the Hanover archives by Count Louis-Alexandre Foucher de Careil (1826–91), a professor at the Sorbonne who was researching Leibniz's work. Foucher de Careil published his discovery of the Leibniz copy of *Olympica* in 1859 under the title *Cogitationes privatae*. Apparently, he understood well that Descartes considered this manuscript, and other papers he left behind in Stockholm, as private. As such, they were written in Latin. Treatises Descartes intended for publication were written in French, to afford them a wide audience in his native country. Descartes, who described his progress through life by saying, "I advance masked," wanted to keep certain things hidden. He had reasons to maintain his secrecy.

According to the inventory made at the time of Descartes' death, *Olympica* (Item 1C) was "a small register in parchment, of which the inside cover bears the inscription: Anno 1619 Kalendis Januarii." But Descartes began to write in the notebook only much later that year, in the month of November, when he wrote, before his night of dreams:

> X. Novembris 1619, cum plenus forem enthousiasmo, & mirabilis scientiae fundamenta reperirem . . .

> [November 10, 1619, as I was filled with enthusiasm, and I discovered the foundations of an admirable science . . .]

What was the great discovery Descartes made on November 10, 1619, which filled him with such enthusiasm? Scholars have contemplated the word "enthusiasm" used by Descartes, looking for hints about the nature of his discovery. Recently, researchers have found a striking similarity between the way Descartes reported his discovery and another one made a few years earlier by the brilliant German astronomer and mathematician Johann Kepler (1571–1630), the man who discovered the laws of planetary motion.

Did Descartes ever meet Kepler? The Kepler scholar Lüder Gäbe has conjectured that such a meeting indeed took place. On February 1, 1620, John-Baptist Hebenstreit, the principal of the high school of Ulm and an associate of Kepler, wrote to Kepler in Linz, inquiring whether

Kepler had received the letters that a certain Cartelius was supposed to have brought him. Hebenstreit wrote: "Cartelius is a man of genuine learning and singular urbanity. I do not wish to burden my friends with ungrateful vagrants, but Cartelius seems a different kind of person, and truly worthy of your consideration."

Gäbe identified Cartelius with Descartes, whose name, in Latin, is Cartesius. In fact, scholars today still refer to Descartes as Cartesius. Kepler's editor Max Caspar noted that a long *s* could have easily been misread as an *l*. Cartesius (Descartes) could well have brought the letters in question to Kepler and made his acquaintance. Gäbe has hypothesized that at some point in his travels, Descartes studied optics with Kepler in Germany.

Whether or not a meeting of the two great mathematicians ever took place, some of Descartes' ideas agree with those of Kepler. Descartes became aware of Kepler's work through his friend Beeckman. He knew all of Kepler's major works; and in Descartes' published work on optics, the *Dioptrique*, which would appear as an appendix to his *Discourse on the Method* in 1637, he would write that Kepler was "my first master in optics."

When Kepler was twenty-three years old—exactly Descartes' age when he wrote the *Olympica* in 1619—he too wrote about a great discovery that filled him with *enthusiasm*. Kepler was looking for a mystical link between ancient Greek mathematics and cosmology. He made what he thought was a stunning connection, and published it in *Mysterium cosmographicum* (1596). Kepler wrote of his rapturous moment of discovery about the planets, again employing a word Descartes would later use, calling it an "admirable example of [God's] wisdom." We know that at some point, Descartes read Kepler's book. Was there a connection between Descartes' own discovery, the one described in his secret notebook, and that of Kepler?

Kepler and his work were associated with a mystical, obscure figure living in southern Germany at that time: the mathematician Johann

Faulhaber (1580–1635). Faulhaber's work on mathematics was of very high quality, but was intertwined with mysticism and the occult. Recently, several scholars have independently analyzed Faulhaber's books, copies of which have been discovered at the Stadtbibliothek Ulm, the municipal library of Ulm, the city in which Faulhaber lived. These researchers have found strong and puzzling connections between Faulhaber's work and Descartes' secret writings. Was Descartes' "admirable science" connected with the work of this mystic-mathematician?

Chapter 5

The Athenians Are Vexed by a Persistent Ancient Plague

IMPELLED FORWARD BY HIS THREE dreams and his interpretation of them, Descartes began to delve deeply into ancient Greek geometry. He spent most of his time alone in his "oven," working out problems and developing ideas. The heart of knowledge was mathematics, but what was the essence of Greek geometry, which Descartes considered the most important part of mathematics? Descartes reviewed the ancient Greek principle of using straightedge and compass to solve all problems. Then he remembered a tantalizing story about Greek construction with straightedge and compass he had heard from his mathematics teachers at La Flèche. This was a story of an ancient mystery whose solution was unknown.

The island of Delos lies at the apparent center of the Cycladic Islands of the Aegean Sea. Since early antiquity, this island has been inhabited, and has always been considered sacred ground. Delos was first settled in the second half of the third millennium B.C. And according to legend, this island is the birthplace of Apollo and Artemis—it has always been the center of worship of Apollo, and became a sanctuary to the god by the seventh century B.C. The Greek city-states of the Aegean competed with one another to build the greatest and most opulent monuments to Apollo. In the seventh century B.C., the Naxians built a terrace with stone lions by the entrance to Delos harbor. The lions can still be seen today, although they are now eroded by millennia of exposure to the sea wind. The island contains countless ruins of ancient temples and shrines. Every city-state in the Aegean Sea had its own temple to Apollo on the island. The Athenians began to influence Delos in 540 B.C. and, after the defeat of the Persians by the Greeks in 479 B.C., founded the Delian Confederacy

A temple on the island of Delos

The Athenians Are Vexed by a Persistent Ancient Plague

of Greek city-states, ostensibly for defense against future Persian invasions, but in reality in order to dominate this coveted island.

In 427 B.C., a plague ravaged Athens, killing a quarter of its population, including the great leader Pericles. In desperation, the Athenians sent a delegation to Delos, to entreat Apollo's oracle to beg the god to spare their lives. The oracle returned with the god's demand: Apollo wanted the Athenians to *double* the size of his temple on the island. The Athenians quickly set to work. They doubled the length, the width, and the height of the Athenian temple to Apollo. They decorated it opulently and lavished it with gifts, and soon the Athenian temple on Delos was the most magnificent on the island, or perhaps anywhere. The delegation returned to Athens with great hope, expecting that the god would now lift the curse. But the plague continued to ravage the city. So a second delegation left Athens for Delos. When its members met with the oracle, he surprised them by saying: "You have not followed Apollo's instructions!" The oracle continued: "You have not doubled the size of the god's temple, as he demanded of you. Go back and do as he had commanded you to do!"

Again, the Athenians set to work. They understood their mistake: they had doubled each of the dimensions of the old temple—the length, the width, and the height—and a calculation they now made showed them that they had actually increased the volume of the temple *eightfold* ($2\times2\times2=8$). Apparently the god wanted the *volume* to be doubled, not the dimensions.

Ancient Greek draftsmanship and geometry were always carried out using only a straightedge and a compass, so the Athenian architects did their best with these two tools. But they failed. As hard as they tried, they could not double the volume of the cubic structure that was Apollo's original temple—or for that matter, double the volume of *any* cube—with straightedge and compass alone.

According to Theon of Smyrna (early second century A.D.), the

Athenian architects went to ask Plato for his help. Plato, who had established the Academy in Athens, in which the best mathematicians of his age worked, enlisted the help of the two great mathematicians Eratosthenes and Eudoxus in trying to solve this difficult problem. Eratosthenes was such a superb mathematician that he had been able to estimate with excellent precision the circumference of the earth by measuring the angle rays of light from the sun made at two different locations separated by a known distance. Eudoxus, on the other hand, was the great genius who could compute areas and volumes using methods that anticipated the calculus, which would only be codiscovered over two thousand years later by Leibniz and Newton. But neither Eratosthenes nor Eudoxus could solve the mystery of doubling the cube with straightedge and compass. Nor could anyone else, however fervently urged by Plato, who was desperate to help his countrymen rid themselves of the plague. Plato, who was not a mathematician himself but was called the Maker of Mathematicians because the best mathematicians studied and worked in his Academy, had such great interest in the cube and in other three-dimensional objects of perfect symmetry that such objects would eventually be named after him.

Why was it impossible to double the size of Apollo's temple? If the volume of the original temple was, say, 1,000 cubic meters (having length, width, and height 10 meters each), then the new volume should be $2 \times 1,000 = 2,000$ cubic meters (and not 8,000 cubic meters, as they had obtained on their first attempt by doubling the length, width, and height to 20 meters each). So in order to *double the cube*—in this case, to obtain a structure with volume 2,000 starting with a temple of volume 1,000—they would need to increase the length, the width, and the height by *the cube root of 2* each. This is so because the cube root of 2, when cubed, gives us 2—the factor needed to multiply the volume. This way, each measurement would have to change from 10 meters to 10×(cube root of 2), or approximately 12.6 meters. It turns out that no finite sequence of operations with straightedge and compass can turn a given length to a number that is the product of that length by the cube root of 2—or the cube root of *any* number that is not a perfect cube. The problem Apollo's oracle gave the Athenians was an impossible one to solve. It is important to note that the new temple had to remain in the shape of a cube: otherwise, simply doubling one of its dimensions, say the length, would have done the trick.

The Greeks of antiquity did not know that the *Delian problem*, as it has come to be known, is mathematically impossible, given their tools. An understanding of this problem would have to wait many centuries. In the meantime, they discovered two other problems they could not solve as well, and today we know that these two problems are also mathematically impossible to solve with straightedge and compass. One was the squaring of the circle—to use straightedge and compass to create a square with the same area as that of a given circle. The other was trisecting an angle—given an angle, to use straightedge and compass to divide it into three equal angles. This problem is solvable in special cases, but a general method—one that would work with any given angle—does not exist.

The ancient Greeks—Pythagoras, Euclid, and other great mathematicians of antiquity—were excellent geometers. But they did not have a well-developed theory of algebra. And algebra is needed in order to understand and properly address complex problems of geometry such as doubling the cube, squaring the circle, and trisecting an angle, collectively called the *three classical problems of antiquity*.

Two millennia after the plague of Athens, Descartes found himself mulling over the Delian problem. He contemplated the cube. What were its properties? What was its secret? Why could the cube not be doubled using straightedge and compass?

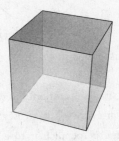

Descartes asked himself the question that was at the heart of Greek geometry—and one that would eventually bring him to an understanding of the Delian problem of doubling the cube, as well as to his great breakthroughs in mathematics: *What does it mean to construct something with straightedge and compass?*

Descartes knew what the two instruments did. The straightedge drew lines and made perfect square corners. The compass drew circles and marked off distances. He asked himself: How do I construct something?

If two points on the plane are given, Descartes (and the ancient Greeks) could construct a line passing through these points. The straightedge is used here.

That was easy enough. Descartes also knew how to construct a cir-

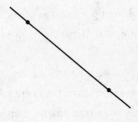

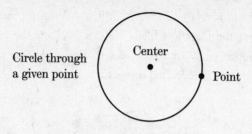

Circle through
a given point

Center

Point

cle centered at one point and passing through another. The compass is used here.

This too was a very simple operation. But more complicated things could also be done. Descartes knew how to use these two ancient instruments to construct a line perpendicular to a given line and passing through a set point. This is how it is done. Use the straightedge to draw the line, and the compass to draw two circles whose intersections are used (again with the straightedge) to draw the line passing through them. This construction is fun.

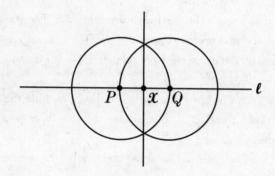

Descartes, and his distant predecessors of ancient Greece, also knew how to construct a line that is *parallel* to a given line and passing through a given point.

Descartes looked at the next figure for a long time and thought. The crisscrossed lines used in the construction were telling. If somehow they could be labeled by their numerical length, then a system could be used

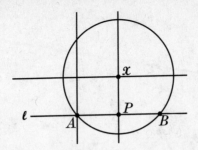

to tie in the numbers with the geometrical constructions. This could potentially allow him to construct a lot more figures than the ancient Greeks were able to create. The use of numbers *and* figures in this way could really unleash the hidden power of mathematics. He would continue to think about how to do it.

Descartes would eventually unify geometry with algebra, bringing an understanding of the three classical problems of antiquity. He would solve several celebrated mathematical problems of ancient Greece, and would also show us the way to solving many more. Descartes' work would shed light on all of mathematics, bringing the wisdom of ancient Greece to our modern world, and would pave the way for the development of mathematics into the twenty-first century. But meanwhile, Descartes also became interested in the mystical aspects of mathematics, and this interest would have strong personal consequences for him.

Chapter 6

The Meeting with Faulhaber and the Battle of Prague

In July 1620, Descartes decided to leave the rest of the troops as they progressed northeast, and to visit the southern German city of Ulm for several months to learn about this part of the country. The first person Descartes met in Ulm was the mystic-mathematician Johann Faulhaber. As well as from Baillet, who gives us this itinerary, there is also a description of Descartes' meeting with this mathematician from Descartes' earlier biographer, Daniel Lipstorp.

Recently studied German documents give incontrovertible evidence that Descartes and Faulhaber did indeed meet. In 1622, Faulhaber published a mathematical text titled *Miracula arithmetica*. In this book, he provided methods for solving quartic (that is, fourth-degree) equations, and these methods are virtually identical to those that Descartes gives in his own book, the *Géométrie*, published in 1637.

In his book published in 1622 Faulhaber wrote:

> Since this noble and very knowledgeable sire, Carolus Zolindius (Polybius), my most favored sire and my friend, has let me know that he will soon publish, in Venice or in Paris, such tables . . .

Faulhaber clearly knew someone named Polybius. And Descartes' hidden manuscript, the *Preambles*, seen and copied by Leibniz as well as studied by Baillet, clearly stated that Descartes planned to write a book about a mathematical truth using the pseudonym Polybius the Cosmopolitan. So Polybius and Descartes were one and the same, and as Faulhaber's book mentions Descartes' pseudonym, this is strong evidence that Descartes and Faulhaber knew each other.

According to Edouard Mehl, a professor of history at the University of Strasbourg who has made a careful study of this issue, Descartes indeed published another book, titled *Thesaurus mathematicus*, under the pen name Polybius. Furthermore, Descartes was traveling to Paris regularly, and while he made it to Venice only in 1624, he had already decided to go there as early as 1620. Thus the claim that Carolus Zolindius (Polybius) planned to publish the mathematical tables mentioned by Faulhaber in either Paris or Venice accords well with Descartes' movements and planned travel during that period. Descartes' works would indeed be published in Paris, and Venice had been an important publishing center ever since the printing presses were invented, and was on Descartes' itinerary.

Mehl concluded that Faulhaber and Descartes were close friends. He argued that "Polybius" was known to Faulhaber as Descartes' pen name, and that Faulhaber was in the habit of calling him by another secret name, Carolus Zolindius, but in his writings also indicated

Descartes' other pen name in parentheses. Dr. Kurt Hawlitschek of Ulm, a leading expert on Faulhaber, further notes in an article about Descartes' meeting with Faulhaber that "Polybius" can be seen to mean "René" (reborn), because the Greek root *poly* means "more," and *bios* means "life." This may have been why Descartes chose this particular pseudonym. When Descartes' secret notebook was analyzed, after Faulhaber's book had come to light, it became evident that Descartes' notebook included, in part, answers to questions raised by Faulhaber.

It is believed that Johann Faulhaber (1580–1635) was born in Ulm and was trained as a weaver. He studied mathematics and was very successful in doing mathematical work. The city of Ulm then appointed him city mathematician and surveyor. In 1600, he also founded his own school in Ulm. Faulhaber's work was in high demand because of his great mathematical skills, and he was often employed in fortification work in Ulm, as well as in Basel, Frankfurt, and other cities. He designed waterwheels and made mathematical and surveying instruments, especially ones with military applications. Faulhaber worked with compasses of various kinds. As mentioned earlier, Faulhaber knew Johann Kepler. He may have worked with him on a number of joint mathematical projects.

Faulhaber studied alchemy, a mystical pseudochemistry whose main purposes were to convert base metals into gold, to find a universal cure for disease, and to discover a way to prolong life indefinitely. He used alchemical and astrological symbols in the work he did in algebra. His work in algebra was extremely important: he studied sums of powers of integers, and his development of results in this area was affirmed by mathematicians who lived after his time.

A further confirmation that Descartes and Faulhaber met comes from the fact that Descartes learned to use the very same symbols Faulhaber had used, symbols commonly found in writings about alchemy and astrology. One of Faulhaber's special symbols appears in Descartes' secret notebook copied by Leibniz. It is the alchemical and astrological sign of Jupiter, shown below.

$$\mathfrak{4}$$

Faulhaber's sign of Jupiter found in Descartes' secret notebook was one of the stumbling blocks to understanding the content of his notebook. No one who studied Leibniz's copy could understand what the sign meant—not until the French scholar Pierre Costabel figured out Leibniz's key to Descartes' mystery. The discovery of Faulhaber's books in Ulm has confirmed that part of the notation used by Descartes had been adapted from Faulhaber's symbols.

We also know that some of the mathematical methods used by Faulhaber were also later employed by Descartes, which makes it virtually certain that the two men indeed met and exchanged mathematical ideas. Both men worked in the same area of algebra: they were interested in extending the work started in the previous century by a quarrelsome group of Italian mathematicians on the solution of cubic equations, that is, equations of the form $ax^3+bx^2+cx+d = 0$.

Faulhaber's interest in mathematics was propelled by his passion for mysticism. He was inspired by the Jewish mystical tradition of Kabbalah. The Kabbalists look at letters of the Hebrew alphabet, attaching to each of them a numerical value (thus aleph=1, beth=2, and so on). By summing the numerical values of all the letters in a word, the Kabbalists seek a hidden meaning by finding other words that have the same numerical sum. The Christian Cabbala is also concerned with numerical values and their symbolism. One key example is the search for the number 666, associated with the beast of the Apocalypse. Revelation 13:18 reads: "This calls for wisdom. If anyone has insight, let him calculate the number of the beast, for it is man's number. His number is 666." Through his very advanced work in mathematics, Faulhaber was searching for significant biblical numbers such as 666. He tried to solve equations and carry out computations that would result in the number 666.

"The first person Descartes met in Ulm was Sire Johann Faulhaber," Baillet tells us, describing the meeting between Descartes and Faulhaber. Descartes came to Faulhaber's house and the mathematician asked him: "Have you spoken of analysis and of the geometers?"

"I have," said Descartes.

"Well, will you be able to solve my problems?"

He handed Descartes a copy of his book. Descartes took the book and looked at the problems of geometry Faulhaber had described in the book. He solved a number of the problems and handed the solutions to his host. Faulhaber laughed. He pointed out harder problems in the book, and Descartes solved those as well.

"Come, now," said Faulhaber. "I want you to enter my study."

As he walked in with his host, Descartes read above the doorway, in German, "Cubic Cossic Pleasure Garden of All Sorts of Beautiful Algebraic Examples." Descartes came into Faulhaber's study, and his host closed the door. Descartes saw bookshelves all around him, overflowing with books. The two men discussed mathematics well into the night, and Faulhaber gave Descartes another book he had written, in German, about algebra. The book was filled with abstract questions with no explanations. Faulhaber asked Descartes for his friendship. Descartes agreed, and Faulhaber said: "I want you to enter a society of work with me." Descartes found that he could not refuse this offer. "Good," said Faulhaber. "Now I would like you to see a book that has been given to me," he said, and he handed Descartes a book written by another German mystic-mathematician, a man named Peter Roth (or Roten). Descartes looked at the problems in Roth's book, and solved them as well. Roth had died a few years earlier. Descartes may not have been aware of it at the time, but Faulhaber and Roth were the two most able mathematicians whose works were associated with a mysterious society so secret that its members were known as "the Invisibles."

Early in November 1620, Descartes and his valet left Ulm and traveled northeast to rejoin the rest of Duke Maximilian's troops, which by now were converging on the city of Prague. Before Descartes had a chance to further study the ideas of ancient Greek geometry and to try to solve the Delian mystery, or fully explore the stimulating problems Faulhaber had posed to him, he was finally called to his first battle. He was eager to see fighting, perhaps as eager as he was to discover truth through science and mathematics.

The forces of Maximilian of Bavaria—who led the German Catholics in this conflict—joined the other armies encircling Prague, all preparing to do battle with the forces of Frederick V, king of Bohemia, defending the city. Descartes and his fellow soldiers were quickly marshaled and made ready for an attack on the city. On November 7, some of the defending forces were able to slip through and regroup on the White Mountain outside the capital. The defenders of Prague had a force of fifteen thousand men, supported by artillery. The attacking force, which included Maximilian's Catholic League as well as imperial troops, was twenty-seven thousand strong. The great battle of Prague had begun.

The defenders enjoyed a quick first victory as their cavalry, supported by cannon, overcame some of the invading troops. But soon the tide turned as the much larger combined powers of the enemy overwhelmed them. By evening on November 8, two thousand men of the Bohemian army lay dead, while the Catholic attackers lost only four hundred. It was clear to the survivors that Prague would soon fall. Frederick V of Bohemia, with his wife, Elizabeth, and their family, hid in the Old Town of Prague, and the king made a hasty plan to smuggle them all out of Bohemia and seek refuge in Silesia.

In the evening, the attacking armies brought their cannon and infantry close to the walls of Prague, after the villages outside the city walls had all surrendered to the Bavarian and imperial armies. On November 9, Descartes and the victorious armies entered the city of Prague. This was the young man's baptism by fire, although Baillet tells us that Descartes did not take part in the actual fighting since he was a volunteer. As Descartes and the other soldiers entered the defeated city, a carriage passed them leaving the city in a hurry. Aboard it was King Frederick V—who would derisively be remembered as "the Winter King" because he lasted only one season—with his family. The departure of the king and queen of Bohemia was so humiliating that they were not even able to take any of their possessions with them. They would be reduced to poverty for the rest of their days and would be treated with contempt both by their enemies and by their erstwhile supporters who had held such high hopes for their reign. Thereafter, only a Habsburg would sit on the throne of Bohemia. One of the members of the fleeing royal family was a two-year-old girl who, like her mother, was also named Elizabeth. Descartes and Princess Elizabeth, unknowingly passing each other in the night, would meet twenty-three years later, and she would become one of the most important people in his life.

Descartes was in Prague with the celebrating troops the next day, November 10, the anniversary of the night of his dreams in the oven, the year before, as well as of his meeting with Beeckman, two years earlier, and of his law thesis, four years earlier. As fate would have it, on this very anniversary of the three watershed events in his life, a fourth event of great importance for Descartes was to take place—most likely inside the city of Prague. While Descartes was walking the streets of this medieval walled city with its ancient towers, majestic bridges on the Vltava River, and magnificent churches, he had a revelation. For it was on the next day that Descartes wrote in the lost *Olympica:* "November 11, 1620. I began to conceive the foundation of an admirable discovery."

What was this discovery? And how was it related to the discovery he began to make a year earlier, in 1619? In her definitive biography, *Descartes*, Geneviève Rodis-Lewis attempts to identify the nature of the achievement about which Descartes is silent. She believes that Descartes' discovery was initiated in 1619 but completed in 1620, and that it is unlikely that he refers here to any of the material he would later include in the *Discourse on the Method* (1637) or its scientific appendixes, since those developments were more involved than any single discovery. His work on unifying algebra and geometry could hardly be traced to a single moment of revelation. Rather, Descartes' rapturous discovery after the heat of battle and the intoxicating influence of victory in Prague led to knowledge he decided to hide—knowledge he inscribed only in his private notebook, written in Latin. The mystical nature of the secret notebook had to have been derived from the influence of Johann Faulhaber on Descartes, especially as reflected in the symbols of alchemy and astrology.

Descartes stayed in Prague until December of that year. The baron of Tilly was left behind in Prague with a garrison of six thousand men, while the rest of the Bavarian troops, led by the duke of Bavaria, left the city. Descartes was there with Maximilian's troops moving to their new winter quarters in the extreme southern part of Bohemia. Six weeks in the capital of Bohemia was enough time for Descartes to learn about the city. While other soldiers pillaged, Descartes found interesting conversations and discussions with the curious and the savants of the city. His greatest pleasure during his stay in Prague was learning from local scholars about the work of the astronomer Tycho Brahe, who had worked in this city, and his greater former assistant, Johann Kepler.

Out of the city in the troops' winter quarters, Descartes once again found himself seeking solitude in his room, spending all his time meditating and studying. He resumed his analysis of geometry, but found himself also mulling his destiny and the road he should take in life. Descartes decided that he wanted to see more action, and more of the world. Staying with the stationary forces holding Bohemia was not to his liking, and at the end of March 1621, he quit the service of Maximilian of Bavaria. Descartes did not want to return to France because he knew that the plague was ravaging Paris; this plague would end only in 1623. So he took his time, traveling north to explore the parts of northern Europe he had not yet seen.

Descartes came back to Holland and visited his friend Beeckman. Isaac Beeckman had had several life-changing experiences in the intervening years. At the end of November 1619 he finally got his position as assistant principal of the Latin School of Utrecht, and five months later, having the stability of a job and an income, had married a woman of Middleburg, on April 20, 1620. Apparently he had been unable to meet someone suitable in Breda, as he had hoped to do two years earlier.

Descartes was happy to see his friend and congratulated him on his marriage. The two men resumed their joint work on mathematics, music, and mechanics. But René confided in Isaac that he had decided to keep some of his own, separate work on mathematics strictly secret. He had reasons for doing so.

Chapter 7

The Brotherhood

During the time Descartes was in Germany and Bohemia, educated Europeans talked about nothing but the emergence in Germany of a secret society of savants known as the Brotherhood of the Rosy Cross. Books purportedly written by members of this society had begun to appear in print just a few years earlier.

Descartes' friends, who knew he was engrossed in science and dedicated to the pursuit of truth, naturally assumed that he was a member of this newly established brotherhood of scholars. According to Baillet, Descartes indeed wanted to get to know the members of this mysterious order devoted to knowledge, and to join their ranks.

"The solitude he endured that winter [of 1620] was always complete, especially with respect to people who were unable to help him progress in his ideas," Baillet tells us. But, he continues, this did not exclude from Descartes' room people who could discuss with him the sciences or bring him news about literature. "It was through conversations with the latter that he learned about a fraternity of savants, which had been established in Germany for some time, under the name of the Brotherhood of the Rosy Cross. His new friends spoke admiringly, but in hushed tones, about this secret society. They told him that the brothers of this fraternity were men who knew everything. They were the masters of every science—they possessed all knowledge, they said: even knowledge that had not yet been divulged."

Descartes saw in these conversations in his "oven" a sign that this was the direction God wanted him to take in order to pursue his destiny of unifying the sciences and searching for knowledge and truth. He yearned to meet these unknown scholars and to join their mysterious organization. Baillet reports that Descartes confided to a friend his view that the Brothers of the Rosy Cross could not be impostors since "it would not be right that they should enjoy a good reputation as possessors of the truth at the expense of the good faith of the people." He decided to make an effort to find them. But here Descartes came up against an insurmountable difficulty. By their very constitution, the Brothers of the Rosy Cross—also known as the Rosicrucians—were unidentifiable. People called them "the Invisibles." They were no different in appearance and habits and customs and everyday behavior from the rest of the population. And their meetings were secret and closed to outsiders.

Despite all his efforts and the inquiries he made to everyone he knew, Descartes was unable to find a single person who would confess membership in the Brotherhood of the Rosy Cross, or who was even suspected of such membership. He was apparently completely unaware that he had already met, and developed a friendship with, one of the most prominent mathematicians associated with the Rosicrucian order—Johann Faulhaber.

The Brotherhood of the Rosy Cross was a secret mystical society of scholars and reformers established in Germany in the early part of the seventeenth century. The symbol of the society was a cross with a single rose in its center.

The story of the establishment of the Brotherhood of the Rosy Cross is truly fantastic. This story is told in the first Rosicrucian text, the *Fama fraternitatis*, published in 1614; it is repeated almost verbatim in Baillet (1691), as well as in several other sources such as Heindel and Heindel (1988). The original founder of the society was a German man of a poor family, but with noble origins, who was born in 1378. His name was Christian Rosenkreuz (in German, "Rosy Cross"), leading to the name of the society.

When he was five years old, the parents of Christian Rosenkreuz placed him in a monastery, where he learned Greek and Latin. At the age of sixteen he left the monastery and joined a group of magicians, learning their art and traveling with them for five years. Then Rosenkreuz left the magicians and continued to travel on his own. He went to Turkey, and from there on to Damascus and farther into Arabia. There he heard about a secret city in the desert, known only to philosophers, whose inhabitants possessed extraordinary knowledge of nature. The city was named Damcar.

Rosenkreuz found his way to Damcar and was received with great hospitality by its citizens. They all seemed to have been expecting his arrival. He described to them his experiences in the monastery and his travel with the band of magicians, and they instructed him in all their knowledge. They shared with him their science and their understanding of the laws of nature, including physics and mathematics.

After three years with the people of Damcar, during which time he absorbed their secret knowledge of the universe, Christian Rosenkreuz left and traveled to the Barbary Coast. He went to the city of Fez, and settled there for two years, meeting sages and cabalists and learning their art. He developed ideas about reforming all of science and reforming society. He went to Spain hoping to spread his new knowledge and ideas to Europe. But the people he met were opposed to his knowledge and his theories, and treated him with contempt. He traveled through Europe, finding no interest in his ideas or his science, only disappointment, opposition, and ridicule. Finally, Rosenkreuz returned to Germany and built a large house and pursued his knowledge in solitary research. He would keep the wonderful science he was uncovering all to himself, rather than seek the glory of recognition by society. He constructed scientific instruments and conducted experiments in his house. He wanted to reform the world using science, and his dream was that after his death, his ideas would be carried forward by a select group of scholars. In the year 1484, Rosenkreuz died naturally with no illness at the age of 106.

Christian Rosenkreuz was buried in a cave that he had fitted with many gold vessels and that seemed to possess magical qualities. In 1604, exactly 120 years after he died, his burial cave was serendipitously discovered by four scholars. The cave had natural light shining in it, even though sunlight could not come in. There was a bright plate of copper with mysterious writings on it, including his initials: R.C. There were four figurines, each inscribed with writings; and there were items that belonged to the deceased: mirrors, bells, books, and an open dictionary.

Everything in the cave sparkled brightly. But the most remarkable thing in the cave was this Latin inscription:

Post CXX Annos Patebo

[After six score years, I shall be found]

The four friends took this as a sign. They learned Rosenkreuz's secrets from the possessions and writings he had left behind, and they decided to found the secret Brotherhood of the Rosenkreuz, the Rosy Cross. The purpose of their order was a general reformation of the world using the sciences. They embraced the study of mathematics and physics, but were also interested in medicine and chemistry.

Within a short time, the four members of the fraternity brought in one friend each, and there were now eight of them. The brothers made the following six rules:

1. They must heal and distribute free medicines to all people who need them.
2. They must dress in accordance with the customs of the country in which they live.
3. They must meet once every year.
4. Each must choose a successor, so that all of them will be replaced once they die.
5. Each must carry a hidden seal with the letters *R.C.*
6. They will keep their society secret for at least one hundred years.

The brothers sought to develop a magical language that would serve as a secret code for science. They dispersed throughout the world, each dressing and acting in accordance with the laws and customs of the land in which he lived. Their mission was to divulge their knowledge and to rectify all the errors of science and society.

In 1614, ten years after it was founded, the brotherhood published its major book, called the *Fama fraternitatis,* or "Statement of the Brotherhood." There followed the *Confessio fraternitatis* ("Confession of the Brotherhood"), in 1615; and a year later, *The Chemical Wedding of Christian Rosenkreuz.* The term "chemical wedding" derives from alchemy, in which chemical elements are to be wedded together to produce gold. We don't know who authored the first two Rosicrucian texts, but scholars have identified the author of the *Chemical Wedding* as the Lutheran theologian Johann Valentin Andreä (1586–1654). The three publications received an enormous amount of attention, and caused great excitement among various groups in Europe, making the brotherhood the talk of the entire Continent during the period Descartes was in Germany a few years after the publications first appeared. According to Baillet, word about the founding of the new Brotherhood of the Rosy Cross spread around the world "like news of a Second Coming." Original copies of the Rosicrucian texts exist today.

An anonymous seventeenth-century work entitled the *Chevalier de l'aigle du pelican ou Rosecroix* describes a ritual at the first Rosicrucian lodge. The brothers are dressed in black sashes and aprons. The master stands in front of a table on which are placed three items: a perfect metal triangle, a compass, and a Bible. The master takes a seven-pointed star and lights its points; the flaming star is passed around. The ritual symbolizes the brotherhood's interest in geometry as well as in the physical world and in religion.

Some people maintain that the Rosicrucians never existed, and that everything said or written about them is sheer myth. According to

Baillet, those who hated the Rosicrucians called them Lutherans, believing that Protestants had invented a fictitious society to foment revolution. But books have survived, attributed to the Brotherhood of the Rosy Cross, and in these books are writings about mathematics, science, and mysticism. Also, these works incorporate a philosophy of life and an approach to politics that are unique and were revolutionary for their time. Clearly, there existed during the early seventeenth century a group of individuals, mostly living in Germany, who knew one another well and called themselves the Brotherhood of the Rosy Cross. If they defined themselves that way—and their books attest to their existence and association—then who are we to say that they never existed? To doubt the existence of the order of the Rosy Cross might not be different from doubting the existence of other secret societies, such as the Pythagorean order of Greece of the fifth century B.C., which gave us the Pythagorean theorem and early ideas about irrational numbers.

The Rosicrucians delved in mysticism, alchemy, and astrology. They studied mathematics and early notions of physics, as well as biology and medicine. The Rosicrucians believed that all knowledge was valuable and that it should be unified and pursued as an entity all its own. Mathematics held a key role in all science and could be used to explain the forces of nature. Hence, their philosophy of science was somewhat akin to that of the Pythagoreans, who considered geometry to be at the highest level of human knowledge. It was also very close to the ideas that Descartes would later express in his writings.

The Rosicrucians were opposed to the power of the church and advocated reform of the religious system on the Continent. The brothers were concerned about the Catholic Church's opposition to scientific ideas, and they sought change. This may well have been one of the main reasons that the Brotherhood of the Rosy Cross was a secret society. Had they not maintained secrecy, they would have been persecuted and severely punished by the Inquisition. Rosicrucian writings indicate that the order was opposed to national loyalty—the members saw themselves as citizens of the world rather than of any country. In addition to the unity of all knowledge, the Rosicrucians advocated a unity of humanity without any national or ethnic boundaries.

Knowledge was traditionally pursued at the university, and in Europe of the early seventeenth century, the universities were dominated by the scholastic tradition and Aristotelian thinking. The Jesuits embraced the Aristotelian view of the universe, which agreed with scripture in placing the earth at the center of all creation. Thus teachings and research at the university were unwelcoming of the new ideas of Copernicus, Kepler, and others who promoted science. Thinking in Europe of this period was institutionalized and was closed to any new ideas or interpretations. The Rosicrucians were opposed to this trend and to the institutions of their time, the church and the universities, and advocated the view that knowledge should be pursued outside these institutions.

Given the political role of the Rosicrucians, it is difficult to maintain skepticism about their existence. In a society in which people are in danger whenever they express views—political, scientific, or religious—that are in opposition to those of the authorities, secret societies and organizations arise. It is known that the Jesuits put a man named Adam Haslmayr in irons on a galley for publishing a treatise in 1614 amplifying the writings in the *Fama fraternitatis*. He was arrested by the Jesuits shortly after the publication appeared. Haslmayr had openly declared that the Jesuits had usurped the title of a true Society of Jesus from its rightful owners—the Rosicrucians. According to Haslmayr, who was a close collaborator of Johann Valentin Andreä, widely believed to be the key Rosicrucian writer, the main purpose of the Brotherhood of the Rosy Cross was to unite the sciences under the canopy of a Christian society. Thus they attracted to their fold historians, linguists, chemists, physicists, and mathematicians. Descartes' nat-

ural secretiveness and his pursuit of science without limits agree well
with stated Rosicrucian beliefs. Was the man who wrote that he "advanced masked through life" a Rosicrucian?

The original legend about the genesis of the Rosicrucians, repeated almost word for word in all the Rosicrucian sources, may well have a
sound factual germ. The legend reflects the transfer of knowledge from
East to West that took place in medieval times. We know that after the
decline of the West following the destruction of Rome in the fifth century, the intellectual center of learning and arts and sciences moved to
Arabia. Here, the ancient texts and ideas of Greece were preserved and
promoted. In the ninth century, a House of Wisdom was established in
Baghdad, in which mathematicians such as Al-Khowarizmi, who gave us
algebra and whose name is reflected in the word "algorithm," worked together with astronomers and other scientists building a new scientific
foundation of knowledge. In the centuries that followed, this knowledge
moved west. The legend about Christian Rozenkreuz bringing such
knowledge from Arabia to Europe may well reflect this historical fact.

The Rosicrucians claimed that they were ancient. In their books,
they said that they were "older than the ancients," in the sense that they
were alive while the ancients existed no more. This claim, which was
propagated through their myth about the ancient origins of their
founder, allowed them to justify their methods. The Rosicrucians argued
that astrology was their predictor of the future and that hundreds of
years of experience with interpretations of celestial signs about human
events allowed them, statistically, to interpret the signs of the heavens
correctly. They made similar claims about alchemy. Thus the Rosicrucians believed that their experience in interpreting the heavens and

chemical reactions allowed them to make correct scientific inferences based on experience. Similar principles allowed the Rosicrucians to claim knowledge of healing the sick. They argued that thousands of years of experimentation with herbs and other medicines enabled them to know which of these ointments and liquids held magical healing powers.

The Rosicrucians claimed ancient origins through the links between their ideas and those of ancient Gnostic and Hermetic traditions. Mystical ideas reflected in Rosicrucian writings originated in alchemical, occult, and mystical writings of third-century Egypt, whose origins were attributed to the much earlier writings of Hermes Trismegistus, believed to have been a contemporary of Moses. The "Hermetic writings" are named after Hermes Trismegistus, but modern scholars believe that these writings go back to a later period, perhaps the second century B.C., and originate in Egyptian magical writings, Jewish mysticism, and Platonism. These elements found their way into seventeenth-century Europe in the writings of the Rosicrucians.

Chapter 8

Swords at Sea
and a Meeting
in the Marais

DESCARTES' THIRST FOR TRUTH IN-
cited his lust for travel. For through visiting new places and seeing how people
lived in different locales, Descartes was
able to free himself of the "false beliefs"
that, he thought, permeated the world
around him. Descartes' emerging philosophy advocated doubt of unproven claims
and finding out the truth through first-
hand observation. For Descartes, such observation was carried out through travel
to distant places and by learning directly
about their people and customs and
lifestyles.

In July 1621, Descartes left Germany
and its invisible Rosicrucians, and traveled to Hungary. At the end of that
month he continued on toward Moravia
and Silesia. Of this part of his voyage, we

know only that he went into the city of Breslaw. The area had been ravaged by the army of the marquis of Jägerndorf, and Descartes was curious about the effects of the hostilities on the inhabitants of this region. Descartes wanted to see more of northern Germany, and to reach the coast. At the beginning of autumn 1621, he found himself in Pomerania, by the Polish border. He found this region to be in great tranquillity and to have very little contact with the outside world, with the exception of the city of Stettin, a major port town with strong commercial links with the rest of the world. Descartes visited the Baltic coast, and then went up to Brandenburg. The Elector, George William, had just returned there from Warsaw and from Prussia, where he had gone to pay his homage to the nobility and his new subjects, as he had just received the title of king of Poland. Descartes continued to the duchy of Mecklenburg, and from there to Holstein.

Before returning to Holland, toward the end of November 1621, Descartes wanted to make one more side trip so he could better observe the Frisian coast and islands. He wanted to travel with ease and agility, so he let go his horses and the rest of his assistants, and kept only his trusted valet with him. He embarked on a vessel on the Elbe, headed for the East Frisian Islands, and planned to continue from the East Frisian to the West Frisian Islands. The Frisians are a group of low-lying islands located in the North Sea off the German and Dutch coasts. The East Frisians belong to Germany, and the West Frisians to Holland. Since Roman times, settlements on these islands have been crumbling away under persistent storms and flooding. Descartes wanted to see some of the partially submerged, abandoned villages on these islands, and to study the problem of trying to keep the sea from reclaiming the land.

Once in the East Frisian Islands, Descartes hired a small boat to take him to the West Frisians, where he wanted to visit a number of specific places that would not otherwise have been accessible to him. According to Baillet, this move "could have been fatal for him." It nearly was.

The crew of this boat was "the roughest and most barbarous lot of their profession." Soon after the boat left port, Descartes realized what a mistake he had made. He saw that the crew of his boat was a pitiless gang of criminals. But he knew that there was little that he could do. Descartes looked wealthy—he was well dressed, a French gentleman with a sword and a bag that the sailors were convinced contained much money, and he traveled with a valet, a sign of prosperity. Descartes tried to keep calm, realizing that hiring this boat and its crew may have been the worst mistake of his life.

The crew, who spoke a Germanic dialect, thought that their passenger, who spoke French quietly to his valet, was a wealthy foreign merchant who certainly did not understand their language. In his presence, they began to talk about their plan to throw him with his valet overboard and make off with his money. Descartes didn't show any sign that he understood anything they were saying as they discussed their cold-blooded plans right in front of him. Baillet explains that there was a difference between highwaymen and these criminals. Robbers on the road, when concealing their identity, would leave their victims unharmed once they robbed them since there was no need to kill them—the victims could not later identify their attackers. But this rough lot had been seen by Descartes. They would have to kill him.

"Surely," they were saying, "this foreigner has no acquaintance in this land, and no one would miss a lone traveler with his valet once the boat returns without them." So they could keep the money without the risk of being caught.

By his polite and calm demeanor, they took him for a weak man who would not give them much of a fight once they attacked him. They went into details of how they would rush the man and his valet, take hold of the two, and—grabbing his bag of money—throw the two men into the freezing water of the North Sea. Descartes kept his sangfroid in the face of this chilling plan.

Having taught himself Flemish a few years earlier, Descartes understood every word of the dialect spoken by these people, and once he knew every detail of their plan, he at once unsheathed his sword and lunged toward them with great force. The stunned sailors backed off, but he pursued them on deck, hurling insults at them in their own language. Demonstrating all his swordsmanship skills and speaking their own language, he swung his sword at them in the air with lightning speed and told them he would cut them all up into pieces. "His daring had a marvelous effect on the spirit of these miserable souls," Baillet tells us. "The terror they felt was followed by vertigo that prevented them from considering the advantage of their number, and they proceeded to conduct him to his destination as peacefully as possible."

Descartes returned from his adventure unharmed and with his money intact. He left the German coast and continued to Holland. He spent the winter in that country, often visiting his friend Beeckman, and also observing with interest the progress of two sieges laid by the Spanish to Dutch cities after five months of truce between the two nations had expired. Then Descartes and his valet traveled through the Catholic parts of the Low Countries, and from there they continued to France. In early 1622, Descartes and his valet passed through Paris and continued south to the regions of Touraine and Poitou. In March, Descartes arrived at his father's house in Rennes.

Descartes stayed with his father for several months. He spent his time riding, socializing, working on problems in geometry, developing his nascent philosophy, and writing furtively in his secret notebook. Since René was now twenty-six years old, his father decided to turn over to him the major part of the wealth he had inherited from his mother after her death, his older brother and sister having taken their shares earlier. Since much of this wealth was in the form of land holdings, René decided to travel to the Poitou region, the location of his new lands, so he could inspect them and perhaps come to a decision as to what to do with them.

Descartes went to Poitou in May of that year, and soon after surveying his vast, sprawling properties, decided to look for buyers. He stayed there until the end of the summer, realizing that such a significant sale would probably take more time to arrange. He knew he didn't want to be a landowner who had to worry about cultivating agricultural tracts and collecting rents from farmers and lodgers, but he wasn't yet exactly what else he wanted to do with his life. He had had quite a ride so far, traveling and seeing the world and participating in and observing battles. He had also made great progress in understanding mathematics and developing a philosophy, but he still was unsure about how to proceed. He returned to his father's house just as fall was beginning to set in, but no one in the family could give him any good advice on what to do next. One thing was sure, however: given the wealth he had inherited—the full extent of which had just been revealed to him on his visit to Poitou—René Descartes would have nothing financial to worry about; he could devote the rest of his life to doing whatever he wanted.

After staying the fall and winter in Rennes, spending time with his sister and brother as well as getting to know his brother-in-law better, René decided in early 1623 to again visit Paris for an extended period of time. He had heard that "a new, clean air was circulating in the capital after three years of contagion from the plague." He was eager to breathe this new air, find new excitement, and renew his old friendships there. He had not seen his friends in Paris for five years.

Descartes moved to Paris with his valet, traveling there on horseback trailed by a caravan of mules carrying his possessions. By then he had found buyers for some of his lands, bought new clothes and items of furniture, and also carried a certain amount of cash with him to deposit in banks in the capital. When Descartes arrived in Paris, the city was awash with stories about the fortunes of war—the very war Descartes had taken part in. Stories were circulating everywhere about the duke of Bavaria; the deposed Frederick of Bohemia, who with his family had by then taken refuge in Holland to live out the rest of their lives in shameful exile; and about the "Bastard of Mansfeld," another military leader of that war—the count of Mansfeld, who was the commander of one of the two armies the Bohemians had put together to face the Austrians and Bavarians.

Knowing that Descartes had participated in that war, and had lived in Germany for some time, many people beseeched him to tell them stories about the action he had seen. To his chagrin, and that of his close friends in the capital, Descartes also discovered that there were rumors about him as well: everyone in Paris was sure that while in Germany, Descartes had joined the secret Brotherhood of the Rosy Cross. It made so much sense to them that he should have done so, being a scientist. And the story on the street was that the secret brotherhood had just sent thirty-six "deputies" to the entire continent of Europe, six of them to France. All six Rosicrucians were "lodged somewhere in the Marais of the Temple in Paris," Baillet tells us. But they could not communicate with the world, and one could not communicate with them, "other than through thought joined with will, that is, in a manner imperceptible to the senses." The chance conjunction of events—the alleged arrival in Paris of six Rosicrucians and the simultaneous arrival of René Descartes—caused the conclusion that Descartes was a Rosicrucian.

In typical fashion, Descartes used his *reason* to combat the assertions he did not like. First, he exploited the fact that the rumored Rosicrucians were "invisible," and he instead made himself very visible. He ensured that he was seen everywhere in Paris—and always surrounded by his many loving friends. Descartes was seen on the streets, at every merry nightspot, and at places where people listened to good music. And second, he stopped doing any mathematics in public. He would study geometry only in the privacy of his own room or—as Baillet, who is our only source on this period in the life of Descartes, tells us—at the request of a friend who might ask him to solve a difficult problem.

Descartes had to be very careful now, for he had apparently begun to use in his mathematical work the mystical symbols from astrology and alchemy learned from Faulhaber. By now, Descartes had understood that his German friend was connected with the Brotherhood of the Rosy Cross. He had to hide all his derivations in his private notebook, for if he was seen developing knowledge using these symbols, it would be impossible for him to defend himself against the charge of delving in the arts of the Rosicrucians. For one thing, the Catholic Church was highly opposed to the Rosicrucians. Were Descartes to be labeled one, his scientific career, and perhaps his safety, could be imperiled

A short time after his return to the French capital, Descartes went to the Marais to visit his college friend Marin Mersenne. After teaching for two years, Mersenne had been elected the *correcteur* (or superior) of his monastery. His confreres soon realized that his talents were far greater than those needed for this job. Consequently, Mersenne was given much free time to study and write. He had a gift for mathematics and science, and the Minim order encouraged him to pursue his studies in these areas, not perceiving any potential conflict between science and faith—something that, perhaps, would not have happened in any other order, the Minims being humble, tolerant, and forward-looking. Marin Mersenne saw his new calling as a conduit of ideas between scientists and theologians. He would foster an understanding among all the major scientists in Europe, and bring them in touch with religious authorities.

Starting out, Mersenne soon realized that a dialogue between the long-accepted scholastic philosophy of the church and the newly emerging scientific revolution was very difficult to maintain, and that often communication between the two camps was in the form of mutual attacks. Mersenne himself began his dialogue with the two groups using a confrontational approach: he attacked the alchemists and the astrologers. But soon he moved to more positive and productive methods of communication. One of Mersenne's former students, Father Jean-François Niceron, moved to Rome to teach at the Minims' house of studies at the Church of the Trinità dei Monti. While in Rome, Niceron made contact with the most influential Italian scientist, Galileo Galilei (1564–1642). The contacts Niceron had made in Italy became very useful for Mersenne. As Mersenne continued building a dialogue between science and faith he began to see his role in life as the director of an international clearinghouse of scientific ideas. He would soon inaugurate what would become known as "the republic of letters." Through letters he would write to and receive from all the major scientists in Europe, Mersenne would establish the model for an international academy of sciences. One of the major players in this drama would be René Descartes. Père Rapin, an ecclesiastic of the time, called Mersenne "Descartes' resident in Paris."

Mersenne's book *Quaestiones celeberrimae in Genesim* ("Celebrated Questions in the Book of Genesis," Paris, 1623) demonstrates how he positioned himself between science and religion. In this book, Mersenne discussed religious topics, but at the same time dedicated forty columns to a description of the laws of optics. After the publication of this book, Mersenne devoted less and less time to religion and spent most of his efforts on science and pure mathematics.

Mersenne soon learned the techniques of printing and became an active publisher. In twenty-five years he produced as many books, totaling over eight thousand pages. Some were his own, while others were written by his scientific correspondents. Mersenne read the major scientific works of his contemporaries: Descartes, Fermat, Desargues,

Roberval, Torricelli, Galileo, and others. But his greatest contribution to science was the connections that he forged as an intermediary between all the major scientists of his day. Mersenne's room in the monastery at the Place Royale was converted into a workshop in which the key scientific and mathematical ideas of the seventeenth century were analyzed and reviewed in the worldwide correspondence he received and sent. Among the most important works analyzed and promoted by Mersenne were Descartes' writings.

Descartes shared with his friend Mersenne his advances in mathematics: derivations of new results based on ancient Greek geometry. When he first came to see him, Father Mersenne was distraught about the new rumors about Descartes. He did not view the Brotherhood of the Rosy Cross favorably, perhaps because he did not consider its members Christians. Mersenne was worried about the consequences Descartes could face if people were to conclude that he was, in fact, a Rosicrucian.

Chapter 9

Descartes and the Rosicrucians

EVEN THOUGH BAILLET DESCRIBED Descartes' interest in, and later denial of any connection with, the Brotherhood of the Rosy Cross, some scholars have persisted in their doubts about the connections between Descartes and the Rosicrucians. But in 2001, Edouard Mehl of the University of Strasbourg published a book, based on his doctoral dissertation from the Sorbonne, in which he analyzed many original sources that had never before been studied. The picture that emerges from his study leaves little doubt that Descartes was deeply influenced by Rosicrucian ideas.

The name Descartes chose for his unpublished notebook, *Olympica*, appears in Rosicrucian writings. And so does the language Descartes used in his *Olympica*.

"Enthusiasm," "admirable science," and "marvelous discovery" had all been used before 1619 as code words by members of the Brotherhood of the Rosy Cross. The name *Olympica* is echoed in at least three treatises on alchemy attributed to the Rosicrucians: *Thesaurinella olympica aurea tripartita* (Frankfurt, 1607), *Rosarium novum olympicum* (Frankfurt, 1606), and in the sentence *"Spiritus olympicus, seu homo invisibilis,"* in the book *Basilica chymica* (Frankfurt, 1620), written by Oswald Croll. Croll was a major writer on alchemy, and used the term "Olympic" to mean *intelligible* or *comprehensible*.

Oswald Croll and Johann Hartmann called themselves the "enthusiasts" of the science of the Rosicrucians in their confrontation with their key detractor, Andreas Libavius, an alchemist who was influenced by mysticism but later came to reject it. Terms such as "marvelous science" and "admirable discovery," as well as other permutations of these words used by Descartes in his *Olympica,* are used in the written exchanges between the "enthusiasts" and their opponent. Croll defined the term "admirable science" as intellectual power and intuition, the image of the Creator in his creatures. He used "admirable science" as code words for philosophy, magic, and alchemy. A coincidence? Perhaps.

But as further evidence that Descartes was familiar with Rosicrucian writings we have Descartes' statement in a letter to Mersenne—"I have faith in no 'sympathetic ointment' of Crollius," a reference to an alchemical universal medicine described by Oswald Croll in his *Basilica chymica.* Croll was the physician of Christian I of Anhalt, who was the counselor of the Elector Palatine, the Winter King, Frederick V of Bohemia (the father of Descartes' future friend Princess Elizabeth). According to the British historian Frances Yates, Frederick V was the man the Rosicrucians had hoped would win the wars against the Catholics and would reestablish Prague as a center for mystical studies and a capital of a realm from which reform of society and religion along Rosicrucian ideals would spread throughout Europe. Yates argues that the Rosicrucians were so closely allied with the prince of the Palatinate that the Rosicrucian text attributed to Johann Valentin Andreä, *The Chemical Wedding of Christian Rosenkreutz*, of 1616, was an allegorical tale, using alchemical symbolism, based on the actual wedding of Frederick and Elizabeth in London in 1613. The Winter King's humiliating defeat in the battle of the White Mountain in 1620 dashed Rosicrucian hopes and eventually led to the decline of the order.

The Rosicrucians published three further major works: a book on spiritual alchemy by Oswald Croll, a treatise on "vitalist philosophy" by Johann Hartmann, and a compendium entitled *Harmonic Philosophy and Magic of the Brotherhood of the Rosy Cross*. The subjects of the three texts, however, are intertwined. In its early years, the brotherhood could be described as a society of alchemists, and its central figure was Johann Hartmann, who held the first European chair in pharmaceutical chemistry, at the University of Marburg, in Germany, in the late 1500s and early 1600s. Hartmann not only published his own books on alchemy, but was also the editor of Croll's *Basilica chymica*. Hartmann is believed to have been in possession of the principal Rosicrucian manuscript, the *Fama fraternitatis*, as early as 1611—three years before the publication of this text.

From Marburg, the center of activity of the Brotherhood of the Rosy Cross moved to Kassel, also in Germany, and the subjects studied by its members expanded from alchemy to other areas. These included theology, botany, astronomy, and mathematics. Maurice of Hesse was an important figure in astronomy, logic, and mathematics, associated with the Rosicrucians. He carried on a correspondence about astronomical matters with the astronomer Tycho Brahe, which was published in Frankfurt in 1596.

The comets of 1618 created much excitement in the population as a whole, and astronomers—some of them reputed Rosicrucians—were especially attracted to this mysterious spectacle in the sky. Brahe's successor, Johann Kepler, first observed the third comet of the year on November 10, 1618 (as did others in Europe)—a date that, as we have seen, figures repeatedly in the life of Descartes.

But the dates November 10 and 11 appeared several times earlier in the history of astronomy. On November 11, 1572, Tycho Brahe first observed a "new star" in the night sky—a supernova, the most dramatic in the history of astronomy since the Chinese observed the supernova that created the Crab Nebula in A.D. 1054. Five years later, in 1577, astronomers at the observatory of Kassel observed on the night of November 10–11 the first appearance of the comet of that year. The observations of this comet made Brahe abandon the theory of solid celestial orbs. This was the ancient Ptolemaic system, named after the second-century mathematician and astronomer Ptolemy of Alexandria, in which the earth is the center of the solar system, and the sun, moon, and planets rotate in concentric spheres around the earth. The comet and its orbit made it necessary for Brahe to modify the ancient Ptolemaic theory of fixed celestial orbs, supported by the church because it agreed with scripture. Brahe did not adopt the complete Copernican model, still maintaining the earth was immobile, but his discovery of the highly elliptic orbit of the comet, which was not aligned with those of the planets, was the beginning of the collapse of the geocentric theory of the universe and constituted evidence in support of Copernicus. The Rosicrucians kept knowledge secret in part because of the implications of their scientific findings on theories the church held sacred.

Perhaps knowing that November 10 and 11 were important recurring dates in the history of science was somehow related to Descartes having his own revelations on that anniversary in 1619 and again in 1620. Remarkably, under the entry "November 11, 1620," Beeckman wrote in his journal about the appearance on that day in 1572 of the comet studied by Brahe, and speculated on the composition of comets as vapors and dust of stars, as well as about their orbits in the sky. He did not mention that this day was the anniversary of his meeting Descartes two years earlier.

Descartes' route from Holland to southern Germany led him through Kassel, and there is a possibility that, at least for a short time, he communicated with the Rosicrucians as early as 1619. According to Edouard Mehl, Descartes encountered members of the Brotherhood of the Rosy Cross in Kassel and found in them a society dedicated to the study of the sciences and mathematics. The unity of scientists of seemingly disparate disciplines may have given Descartes his idea of a "universal science" unifying all knowledge by means of mathematics. Mehl also points out a curious coincidence between Descartes' dreams in the oven and Rosicrucian philosophy. In Descartes' dream, he sees the poem "*Est et Non.*" Mehl points out that a key tenet in Rosicrucian philosophy is the existence versus nonexistence of every element in the universe. The Rosicrucians named their principle *Est, Non est.* Descartes' dream in which he sees his room fill with sparkles of light is reminiscent of the Rosicrucian story about the discovery of the burial cave of the founder of the order, Christian Rosenkreuz. The cave, too, sparkled with light that seemed to come out of nowhere, and the dream itself is similar in its feel to the description of the discovery of this cave. Descartes' dream in which he sees a dictionary bears similarity to descriptions of Rosicrucian rituals in which, too, a dictionary or encyclopedia is used.

Kepler himself was interested in the occult. Was Kepler, too, a Rosicrucian? His assistant, the mathematician Jost Bürgi (1552–1632), whom Brahe described as "the new Archimedes," was a member of the brotherhood. Bürgi invented many scientific and mathematical instruments, including a proportional compass that is very closely related to the one Descartes is reported to have invented, which lends more support to the theory that Descartes communicated with the Rosicrucians. Bürgi's proportional compass, in turn, was a variation of a compass first invented by Galileo in 1606. This was a compass that could maintain set proportions of various quantities because its arms were marked with graduated scales. Galileo had enjoyed some financial success by selling his compass for use in engineering and for military purposes. Bürgi's variation of the proportional compass is shown below.

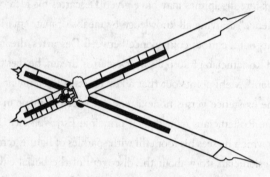

The Bürgi compass

Kenneth L. Manders of the University of Pittsburgh has studied the designs of compasses invented by Descartes and those of compasses described by Johann Faulhaber. According to Manders's interpretation of a particular Descartes text copied by Leibniz and then recopied in the nineteenth century by Count Louis-Alexandre Foucher de Careil of the Sorbonne, Descartes invented *four* kinds of compasses. Manders makes a striking observation. On December 19, 1620, Faulhaber advertised his skills in Ulm in an effort to attract students and obtain contracts for consulting jobs. A published description of Faulhaber's qualifications included the following:

In particular, four new proportional compasses, with which to find geometrically two mean proportionals between two given lines; just so, how to divide any angle on a circle geometrically into three equal parts; equally, how to obtain conic and cylindrical sections geometrically, a matter on which other authors have written large books. Moreover, how to show general rules for equations of arbitrary degree from the number 666.

Manders points out that Faulhaber's compasses are identical to Descartes' compasses, and since these are extremely unusual, special-purpose devices, it is certain that Descartes and Faulhaber met and exchanged very precise information. One of the compasses of Descartes (and of Faulhaber) can be used to solve one of the three classical problems of antiquity: trisecting an angle. But it should be noted that in the original Greek problem, one is to trisect an arbitrary angle using only a straightedge and a simple compass, which was available to the ancient Greeks, not the advanced device invented by Descartes, Faulhaber, Bürgi, or any of them in collaboration. Note that, as also evidenced from Faulhaber's other works, his main interest in mathematics was motivated by his obsession with "biblical numbers" such as 666.

These interests make it almost certain that Faulhaber was indeed a Rosicrucian. In fact, in his book *The Dream of Descartes*, the French scholar Jacques Maritain says the following:

Now Faulhaber was a real Rosicrucian and a very ardent one, and one is justified in assuming, in spite of Baillet's denials, that Descartes found in him the man he was seeking, and that through him Descartes entered into direct contact with the intellectual atmosphere of the Rosicrucians. Might not such contact, however fleeting, have a deciding influence upon the moral lines and the aims of the philosopher's life? May we not even ask ourselves whether at its origin, Descartes' great idea did not permit the supposition that he intended—an intention that became hazier as time went on—fearlessly to transpose to the plane of everyday reason and of the most widespread common sense the design followed on the plane of alchemic mysteries by the naïve Rosicrucians, and in so doing to render it much less "uplifted," but much more efficacious—mathematics replacing the Cabala leading to universal knowledge, the hermetic sciences and their occult qualities giving place henceforth to geometric physics and the art of mechanics, as the elixir of life to the laws of rational medicine?

According to the world's leading expert on Johann Faulhaber and his work, Kurt Hawlitschek of Ulm, Faulhaber was clearly a Rosicrucian. Hawlitschek hypothesizes in his book on Faulhaber that the meeting between Descartes and Faulhaber did not take place by chance. When Descartes was in Frankfurt to witness the coronation of the emperor, before proceeding to join Duke Maximilian's army, he met Count Philipp of Hesse-Butzbach (1581–1643). The count was interested in mathematics and also had Rosicrucian connections. Philipp sent Descartes to Ulm, which was close to where Duke Maximilian's army was encamped, with the purpose of having him meet Faulhaber so they could discuss mathematics.

Two other mathematicians belonging to the Brotherhood of the Rosy Cross were involved in astronomical calculations and the invention and use of compasses. They were Benjamin Bramer (1588–1652) and Faulhaber's friend the mathematician Peter Roth. Both Bramer and Roth are mentioned in Descartes' secret notebook. Descartes most likely met them in Kassel when he traveled to Germany in 1619, and was likely inspired by their work on proportional compasses, which led him to invent his own, related devices.

Faulhaber was the author of a treatise titled *Numerus figuratus sive arithmetica arte mirabili inaudita nova constans*, published in Germany in 1614. When this text was analyzed, after Pierre Costabel had deciphered Leibniz's copy of Descartes' secret notebook, striking similarities in the content of the two manuscripts were discovered. Did Descartes share his "admirable discovery" with his mystic friend Faulhaber? Or was Descartes influenced in his discoveries by the work of the latter? Either way, Faulhaber had at least some knowledge of Descartes' most profound secret.

The Rosicrucians and other mystics were looking for hidden meanings in numbers and shapes: arithmetic and geometry. Descartes was an expert in both areas. He, too, was on a search for meaning and understanding to be found in the realm of numbers and shapes. And the assumption that Descartes was inspired in his secret search by the work of Faulhaber would shed light on the meaning of his writings in the secret notebook. Hiding its content by using mystical symbols assured Descartes that no one uninitiated could understand the writings in his notebook, should it ever come to light.

In January 1618, Faulhaber mildly denied a connection with the Brotherhood of the Rosy Cross, saying: "I will not spare my zeal for obtaining the best information about the very dignified society of the Rosy Cross, but I believe that divine will has not yet determined that I should be worthy of making their acquaintance." But this vow was soon broken when in July of that year he made the acquaintance of several members of the brotherhood, among them an important officer in the order, Daniel Mögling.

Edouard Mehl claims in his book that Faulhaber's mystical mathematical text, the *Miracula arithmetica* of 1622, was published with the help of the Rosicrucians, among them Daniel Mögling. In this book, Faulhaber writes: "It is as impossible to separate the breadth of divine power from the number 666 as it is to separate this divine power from the holy evangelist."

Daniel Mögling even lodged for a time in Faulhaber's house and was influenced by his host's mathematical work. He then changed the focus of his activities from medicine and alchemy to mathematics and astronomy and became the brotherhood's person in charge of these disciplines. Mögling himself had close connections with Johann Kepler. These included a correspondence on astronomical and mystical matters, as well as joint scientific work. Kepler's astrological work, his friendship with Daniel Mögling, and his stated desire for universal reform—echoing Rosicrucian writings—all make it clear that he had at least some association with the Brotherhood of the Rosy Cross.

Faulhaber was proud of having been able to predict, as early as 1617, the future appearance of a comet in 1618. He claimed to have made the prediction based on numerology and Cabbala. Faulhaber computed an astronomical table, and was looking for the apocalyptic number 666. He noticed in his table that the longitude of Mars and the latitude of the moon would both be 3 degrees, 33 minutes, on September 11, 1618. Then he knew that he had found what he was looking for, since $666 = 333 + 333$. This meant to him that something important would appear in the sky on September 11, 1618. So Faulhaber predicted that a comet would be seen on that exact date. In fact, three comets were seen that year, the first one appearing in mid-October. When Faulhaber tried to publish his prediction, *after* the appearance of the first comet, he was accused of having used Kepler's mathematical tables for this purpose, rather than his own numerology.

This quarrel led to a public confrontation in Ulm on October 18, 1619, between Faulhaber and John-Baptist Hebenstreit, the principal of the high school of Ulm, who was an associate of Kepler. Hebenstreit submitted to Faulhaber eight questions about his use of biblical verbs as numbers; and finally, he concluded that Faulhaber's answers were unsatisfactory and blamed him for "mixing the heavens with the earth." Hebenstreit summed up his tirade by accusing Faulhaber of "cabalistic log-arithmo-geometro-mantica." This, in fact, was also the title of a short treatise he published.

Kepler himself never got involved in this dispute, finding it beneath his dignity. He harbored no animosity toward Faulhaber, and perhaps shared a kinship with his fellow mathematician and mystic. Kepler and Faulhaber shared a secret. They were among very few people who knew that Daniel Mögling was the author, using the pseudonym Theophilus Schweighart, of an important Rosicrucian text, the *Speculum sophicum* (1618). Hebenstreit held a grudge against Mögling, and after his attack on Faulhaber had failed, in part because of Kepler's noncooperation, turned his venom against Mögling, attacking him publicly in a very personal, insulting way.

One of Hebenstreit's associates wrote a treatise condemning the Rosicrucians and Faulhaber, titled *Kanones pueriles*, which was purportedly authored by Kleopas Herenius. This name is nothing but an anagram of Kepler's name in Latin: Iohanes Keplerus. Mögling himself, like many Rosicrucians, was fond of anagrams, for they worked well as a device for hiding identities. In 1625, Mögling published a book about perpetual motion, titled *Perpetuum mobile*, which he authored using an anagram of his first name in Latin, Danielis. The jumbled-up letters of "Danielis" gave him the name Saledini, its resemblance to "Saladin" lending it an Eastern, anti-Crusader flavor. This book was found in the inventory of Isaac Beeckman's library, which was part of his journal. Given the fact that Beeckman discussed most of his research and readings with Descartes, the latter likely was aware of Mögling's book.

Descartes' published works, and his letters, make it clear that he was influenced by the achievements and ideas of the two mathematicians Faulhaber and Kepler. Kepler's discoveries about the nature of the orbits of the planets in the solar system were of great interest to Descartes and, according to some scholars, were at the heart of his "admirable science." And Descartes' secret notebook, by its unique use of symbols and by its content, echoed the writings of Faulhaber. Descartes' writings testify to the fact that he had at least exchanged ideas with the Rosicrucians.

Chapter 10

Italian Creations

AFTER TWO MONTHS AND A FEW
days in Paris, a period during which he pre-
tended to all but his confidant Mersenne
that he had renounced the study of mathematics—to avoid the trap of being labeled a Rosicrucian—Descartes left for
Rennes. He arrived there at the beginning
of May 1623. From there, René went to
Poitou and stayed until July. He managed
to sell, with the consent of his father,
most of his holdings in this region. On
July 8, 1623, Descartes sold a large estate
he inherited from his grandparents, called
"the land of Perron," to Abel de Couhé, a
nobleman of Poitou, and the sale was
sealed by the notaries of Châtellerault. He
took the cash with him to Brittany, and
after bidding his family good-bye, returned to Paris.

Descartes was unable to decide how to use all the money he had brought with him from the countryside. Much of it went into a bank account; he apparently wanted to place some of the funds in investments, but found none to his liking. He decided to use some of his new money to support a long trip to Italy.

Descartes wrote a good-bye letter to his father, saying "A voyage beyond the Alps would be of great use to me, to instruct me in handling my affairs, acquire some experience in the world, and form new habits that I do not yet have. If I will not become richer, at least I may become more capable." Perhaps the young man felt he needed to justify to his father taking an expensive trip to be paid for by what was family money, while before that time, when he was at least nominally associated with an army, he could explain the travel as part of his occupation.

Descartes crossed the Alps and continued east to Zurich. He walked down the wide, cobblestoned Neumarkt Street with its medieval mansions and the towering church at the center of the old town. He sought out scholars and savants, and finding them, discussed nature and mathematics.

Descartes continued east to Tirol, and from there he descended to the plains of northern Italy, arriving in Venice just in time to witness the ceremony of the wedding of Venice with the sea on Ascension Day. Descartes arrived at the Church of San Nicolò on the Lido to view the *bucentoro*, a specially outfitted, gilded galley carrying the doge as he was rowed out to sea from the Port of San Nicolò. Descartes sat close to Venice's leading families and envoys from foreign states who had come to witness this unique ceremony. When the galley had traveled some distance from the port, the doge cast a golden ring into the waters of the Adriatic Sea and proclaimed by this act that he was taking dominion of the sea as a husband over his wife.

Legend has it that the pope had given a ring, and with it lordship over the Adriatic, to the doge in 1177. According to this legend, the Venetians defeated an imperial fleet, following which the Holy Roman

Emperor Frederick I Barbarossa (Italian for "Red Beard") came to Venice to kiss the pope's feet. However, the victory and the battle never took place, and were pure fiction. But the ceremony of the wedding of Venice with the sea was celebrated nevertheless, and four and a half centuries after this custom had begun, Descartes witnessed it.

From Venice, Descartes headed south to Rome—a city he had always longed to see, since it was the heart of the Catholic world and an exciting, vibrant cultural center. He passed through Loreto and fulfilled his vow to visit this religious shrine. After completing his visit to Rome, he was ready to return to France, but decided to stop in Tuscany. Descartes had heard much about Galileo and admired him greatly. He was hoping to meet him at his home in Arcetri. Descartes made it to Tuscany—but to his great disappointment was never able to meet Galileo.

April 1624 found Descartes in the town of Gavi, observing the military maneuvers of the duke of Savoy. Then in May he went to visit the city of Turin. Descartes climbed up into the Alps that rise over the border region between France and Italy. He spent some time in the mountains observing the melting snows and taking note of how thunderstorms developed. During his Italian trip, Descartes also observed rainbows and parhelic circles: luminous circles or halos parallel to the horizon at the altitude of the sun. These "false suns" appeared in Rome while Descartes was visiting that city and caused great excitement in the population. Everyone wanted to know how this phenomenon developed. Years later, when his *Discourse on the Method* and its scientific appendixes were published, Descartes gave his explanations of the natural phenomena he had observed on his trip to Italy. Descartes' Italian trip was an important event in his life since it allowed him to study nature. But there were other benefits to this voyage as well. Through his travels in Italy, the young man matured and developed a clearer sense of who he was and of the scholarly goals he would pursue in his life.

Descartes continued to Rennes, enjoyed the company of his family, and then moved back to Paris. Throughout this period, Descartes was working on algebraic problems studied by Italian mathematicians who had lived during the century before his time in the plains of the Veneto—the region of Venice, which he had just visited.

The Babylonians understood simple concepts of equations: given a specified set of arithmetical conditions, they could solve an equation to obtain the value of an unknown quantity. For example, they knew how to find the length and width of a field that was to have a specified area, given some condition on the two measurements. The ancient Egyptians also knew how to solve such problems. Papyri that survive, for example the famous Ahmes papyrus, now kept at the British Museum and dated to approximately 1650 B.C., show the solution of simple equations. For example, problem 24 in the Ahmes papyrus asks for the value of a "heap," if a heap added to a seventh of a heap gives 19. Ahmes gives the answer as $16+1/2+1/8$, which is 16.625, as we would obtain today by solving the equation $x+(1/7)x=19$.

The ancient Greeks were also able to solve equations, so that by the close of the classical period of ancient Greece and Rome, people knew how to solve some quadratic equations, that is, equations with highest-order term being ax^2. But no general method had been known for solving such equations, nor was anyone able to solve higher-order equations (equations with powers higher than 2—for example, equations containing a term ax^3).

The word "algebra" comes from the first two words in the Arabic title of a book written about A.D. 825 in Baghdad: *Al-Jabr wa-al-Muqabalah*, by Muhammad ibn Musa Al-Khowarizmi. In his book—the first important book on algebra—Al-Khowarizmi presented complete methods of solution for the quadratic equation. In Descartes' notation as we use it today, such equations are written in this general form—

$$ax^2+bx+c=0$$

—and every high school mathematician knows the general formula giving the two roots, or solutions, of this equation.

While people knew how to solve quadratic equations, no one knew how to solve a *third-order* equation:

$$ax^3+bx^2+cx+d=0$$

How to solve such an equation remained a mystery for another seven centuries beyond Al-Khowarizmi—until four Italian mathematicians, working one against the other, endeavored to solve it.

Who were these Italian developers of algebra? They were four mathematicians who lived in the sixteenth century in northern Italy—the same area Descartes visited in his Italian trip in 1623–24. And they were a somewhat shady and unsavory group of individuals.

Niccolò Fontana (1499–1557), known as Tartaglia ("the Stammerer," in Italian) was born in Brescia, in the republic of Venice, in 1499. As a fatherless boy of thirteen, he almost died in 1512, when French forces looted his hometown and many people were killed. The boy was dealt severe facial wounds from a saber that cut his jaw and palate, and he was left for dead. His mother found him and nursed him back to health. As an adult he grew a beard to hide his scars, and he could speak only with difficulty, hence the nickname Tartaglia.

Tartaglia taught himself mathematics. Having an extraordinary ability, he earned his living teaching mathematics in Verona and

Venice. As a mathematics teacher in Venice, Tartaglia gradually acquired a reputation by participating successfully in many public competitions.

The first person in history known to have succeeded in solving cubic equations was Scipione del Ferro (1465–1526), who was a professor of mathematics at the University of Bologna. Nobody knows how del Ferro made his amazing discovery. He did not publish it and did not divulge it to anyone—until he was on his deathbed. Then he passed the secret on to his mediocre student Antonio Maria Fior. Soon afterward, word was out that Fior could solve cubic equations, which was considered a great achievement since no one had been able to do so even though people had been trying for many centuries.

In 1535, Fior challenged Tartaglia to a competition. Each person was to submit thirty problems for the other to solve. At that time, the person who won such a competition could expect money, prestige, and sometimes a professorship at a university. Fior was confident that his ability to solve cubic equations would be enough to defeat Tartaglia, but del Ferro had shown Fior how to solve only one type of equation: a cubic equation of the simple form: $x^3=ax+b$ (that is, the coefficient of x^3 is 1, and there is no x^2 term). Tartaglia, on the other hand, submitted to Fior a variety of different problems. Fior's poor performance, since he had only the one recipe given him by del Ferro, exposed him as an inferior mathematician. Fior also gave Tartaglia thirty problems of a variety of kinds—being sure that he could not solve them since he himself didn't know how to do so. Unbeknownst to him, in the early hours of February 13, 1535, Tartaglia had a revelation: he discovered a *general* method of solving cubic equations, that is, equations of the very general type: $ax^3+bx^2+cx+d=0$. Tartaglia was now able to quickly solve all thirty of Fior's problems—he completed his task in less than two hours. It was clear to everyone present that Tartaglia was both the winner of the competition and the greatest mathematician.

At this point we meet another Italian mathematician, Girolamo

Cardano (1501–76), who was a medical doctor and mathematics lecturer at the Piatti Foundation in Milan. Cardano was well aware of the importance of solving cubic equations, and was intrigued when he heard about the results of the contest in Venice. He immediately set to work on trying to discover Tartaglia's secret, but he was unsuccessful. A few years later, in 1539, he contacted Tartaglia through an intermediary. Cardano told Trataglia that he wanted to include his method for solving cubic equations in a book he was planning to publish that year. Tartaglia declined the offer, saying that he wanted to publish his own book. Cardano then asked Tartaglia if he would mind showing him his method anyway, promising that he would keep it secret. Tartaglia again refused.

Not giving up, Cardano now wrote Tartaglia a letter in which he hinted that he had been discussing Tartaglia's brilliance with the chief of the army in Milan, Alfonso d'Avalos, who was one of Cardano's powerful sponsors. Tartaglia took the bait. He was a poorly paid mathematics teacher, and the thought of meeting an influential and wealthy individual who might be able to help him appealed to him. He wrote back to Cardano, and Cardano then invited him to his house, promising that he would arrange a meeting between Tartaglia and d'Avalos.

On March 25, 1539, Tartaglia left Venice for Milan. To his dismay, he found that d'Avalos was not there as Cardano had promised. Cardano, however, wined and dined him in his house, and tried every method he could to convince Tartaglia to reveal his secret. Late that night, after he had drunk much wine and Cardano had sworn to him that he would never reveal his secret, Tartaglia divulged to him his secret formula. He did it by way of a poem in Italian in which he embedded his formula.

In 1545, Cardano published his now-famous book *Ars magna* ("The Great Art"), which contained solutions to the cubic equation based on Tartaglia's secret formula, as well as solutions to the quartic (fourth-order) equations, which had been obtained by Cardano's student

Ludovico Ferrari (1522–65). Cardano thanked Tartaglia in his book. But he had broken his promise—the oath he swore to Tartaglia never to reveal his secret. Understandably, Tartaglia was furious and for years kept writing letters to everyone he knew, attacking Cardano. He even published their conversation in Milan and the broken promise, including the formula he had divulged. But Cardano's book, the *Ars magna*, had established him as a leading mathematician, and he was untouched by Tartaglia's attacks. To add to his misery, Tartaglia was never given the chance to meet the wealthy patron he had hoped would help him. After a short period of teaching at a university, he returned to his position as teacher in Venice, which he kept until his death.

Today, Tartaglia is remembered together with Cardano for a formula for solving cubic equations. Tartaglia also wrote a popular arithmetic text, and was the first Italian translator and publisher of Euclid's *Elements*, in 1543. He also published Latin editions of Archimedes' works.

Descartes was well aware of the genesis of algebra and the development of solutions to equations of third and fourth order. He spent time working on such problems, and early on derived a result in this area. Descartes had shown that if a quartic equation has a special form (has no cubic term) and can be factored into two quadratic equations—

$$x^4+px^2+qx+r=(x^2+ax+b)(x^2+cx+d)$$

—then the number a^2 is the root of a cubic equation; and also, b, c, and d are then rational numbers (meaning fractions or integers) that depend on a. This is a useful result that sometimes helps solve equations. It was a good start, and a continuation by Descartes of the work done a century earlier by the Italians. It was also strikingly close to work done by Faulhaber.

The Italian mathematicians, the early algebraists, were called *cossists*. The word "cossist" comes from the Italian *cosa*, meaning "thing." The *cosa* was the mystery that algebra was designed to solve—it was the name given to the unknown quantity in an equation (our modern *x*). Descartes' missing notebook contained an original alchemical and astrological sign, the sign of Jupiter. But it also included early cossist notation:

Descartes learned to use this sign from the Italian cossists, whose algebra he had studied. Descartes himself later invented our modern notation, used in algebra today. He taught us the use of *x* and *y* for variables to be solved, and *a* and *b* and *c*, and so forth, for known quantities. But tantalizingly, his hidden notebook used a different notation—the mystical notation inspired by alchemy, astrology, and Rosicrucianism, and the old cossist notation. The secret notebook contained yet a third sign, which no one has been able to trace to any previous source.

Incidentally, Descartes never used the equal sign (=), even though it had been invented by Michel Recorde in 1557. Descartes persisted in his use of a backward-facing Greek alpha ✂ to denote equality. Interestingly, it would be Leibniz who decades after Descartes' death would revive the use of the equal sign we use today.

Chapter 11

A Duel at Orléans, and the Siege of La Rochelle

DESCARTES TRAVELED ONCE MORE to Touraine and Poitou, and then went back to the capital and settled there for a few months. But family issues in Touraine, Poitou, and in Rennes kept him traveling to these areas frequently. He spent time with his sister and his brother-in-law, and with his father. René traveled to Poitou to sell more of his lands and to liquidate other assets so he could take the money to Paris and live comfortably on the wealth he had inherited. He also traveled frequently to Touraine to visit his governess and family members who had remained there.

When he was just becoming a young adult, some years earlier, Descartes was rumored to have had an intimate relationship with a mysterious woman of Touraine called La Menaudière. His family, disapproving of his relationship with someone they considered unsuitable for their son, began to look to find him a wife. They thought that getting married might bring stability to the life of the young and restless adventurer, and that once he married, he might settle down in the area and establish a business. The search began in earnest to find René a wife "of good birth and much merit."

There was a very beautiful young woman, who later became known as Mme. de Rosay, whose family also lived in Touraine. She was indeed of good birth and suitable to the Descartes family. René and the young woman met several times, and they were attracted to each other. But René soon left on his travels, and their relationship never progressed.

But now, in 1625, René was again traveling the roads of Touraine and the French countryside south of Paris to see his family and take care of his affairs. He hadn't seen her for a few years. On one of his trips between La Haye and Paris, Descartes stopped his horse-drawn carriage at a major intersection near the city of Orléans, which lies on the main road south from Paris. Descartes had to stop for a while to let his horses rest, eat, and drink. He entered a roadside inn, one that was popular with travelers on these routes. France had many such establishments along its highways. The inn had a closed courtyard in which the horses were sheltered and fed and watered. Inside the inn there was a large room with a ceiling supported by dark wooden beams. It had large, arched windows. There were several wooden tables surrounded by simple rustic chairs. People crowded around these tables drinking and eating.

Descartes and his valet spent some time at the inn, enjoyed a meal and rested, and were ready to continue their journey north to Paris. They walked into the courtyard, collected their horses, and went into the sunshine. They harnessed the horses to the carriage and were just about ready to leave when Descartes suddenly looked up. And there she was: the woman we now know as Mme. de Rosay.

The two people looked at each other, and it seemed that the intervening years had done nothing to cool the attraction between them. Descartes, dressed in green taffeta and looking very smashing with his plumed hat and sword, approached her. She looked right into his eyes. They stood there speechless for a moment, just staring at each other. And then her companion rushed over. He was a jealous man, and he unsheathed his sword and challenged Descartes to a duel.

The man apparently didn't know whom he was dealing with—someone with much experience with swords and battles. The two men locked swords, and swung and parried for a few moments. Swiftly, Descartes brought his sword in one last time and delivered a final blow. His opponent's sword flew up into the air. Descartes put the point of his sword to his challenger's throat and, glancing at Mme. de Rosay, said to him: "The lady has beautiful eyes, and for that I will spare your life." He let him go, and pulled back in disgust. The lady rushed over to Descartes' side. One last time, Descartes stared into those beautiful eyes, and turning away from her, he said: "Your beauty is unmatched, but I love truth the most." He left the two stunned figures by the roadside and in a minute gathered his valet, and in a whirl of dust they were off to Paris.

Years later, when she was married and had become Mme. de Rosay, and when Descartes had become a famous philosopher, the lady confessed the story to her priest, who in order to protect his identity when he repeated the tale—violating the secrecy of the confessional—remained known only as "Father P."

According to Mme. de Rosay, she saw René Descartes for the first time when he was a young man and was one day in the company of several other young men who were joyfully playing around and talking about women. He confessed to them that he had never yet met a woman he found irresistible. Then he said: "I find that a beautiful woman, a good book, and a perfect preacher are the three most difficult things to find in this world."

But then, according to the lady, he met her. And he found her to be beautiful and irresistible, and from then on he desired only her. "René Descartes was a young cavalier who was guided by a love for me," she said. "It made him distinguish himself in great deeds on my behalf."

According to Mme. de Rosay, Descartes was accompanying her, along with other ladies, on a trip to Paris when they were approached at Orléans by Descartes' bitter rival for her affections. When Descartes had won the duel and put his sword to his rival's neck, he told him the following: "You owe your life to this beautiful woman, to whom I devote my own." But alas, things stopped there, and Descartes never married the lady. Was her version of the story true? Or did Descartes really say that he loved truth more than her beauty? Most likely, Descartes' account was the correct one, since it is in line with his general behavior and his rather cool approach to relationships. And we know that Descartes loved beautiful eyes.

His frequent trips to Touraine and Poitou over, René Descartes was living in Paris in 1628, hiding from his friends once again so he could work in peace. He was writing extensively, deriving important results—some of which would become public within nine years. This was a difficult period for Descartes because he was anxious about not being found by his many friends as well as strangers who were attracted to him because of his growing fame as a philosopher, scientist, and mathematician. "The displeasure he felt by having been chased out of his favorite quarter into hiding brought about in him a desire to go to see the siege of La Rochelle," Baillet tells us.

Of all the places in which the Huguenots were once safe, by the early seventeenth century there remained only the city of La Rochelle, on the Atlantic coast of France. La Rochelle is a medieval walled city built in the twelfth century. It has a fortified harbor with two imposing ancient crenellated towers guarding its entrance from the sea. There is a fifteenth-century tower called the Tour Lanterne, which served as a powerful lighthouse guiding merchant vessels from around the world to this prosperous city. La Rochelle came to prominence with an economy based on salt, wheat, and wine. By 1620, 85 percent of its population was Huguenot. But the French state, whose religion was Catholicism, was threatened by the power of the Huguenots. King Louis XIII and his minister, Cardinal Richelieu, made a decision to crush Huguenot power.

In 1627, the people of La Rochelle requested and obtained the aid of the British fleet against the French. King Louis XIII then sent his forces to La Rochelle in response to this act, to confront the British fleet, which made its base on the nearby Isle of Ré. The French army wanted to prevent the British forces from aiding the Huguenots. The king himself ordered and oversaw a siege of the city—in what became one of the most dramatically recorded conflicts in French history. Alexandre Dumas devoted part of his masterpiece *The Three Musketeers* to the siege, sending Athos, Pathos, Aramis, and d'Artagnan to the scene. Cardinal Richelieu had his headquarters outside the city, commanding half the French forces, while the king commanded the other half.

The siege of La Rochelle lasted thirteen months, from September 1627 until the city fell in October 1628. As the siege began, French forces encircled the city from all sides, preventing food and supplies from the rich lands it owned in the countryside from reaching the walled city. But there still remained the heavily fortified harbor, and through it the British were able to use their ships to bring in supplies of food to the besieged city.

Soon after the siege began, the French decided to build a dike across the entrance to the inlet leading to the harbor. The aim of the French forces was to block completely the entrance to the harbor and thus prevent the British fleet from entering it. In March 1628, the king left for Paris and made Cardinal Richelieu the lieutenant general of the army.

The cardinal began to construct the dike across the inlet, and when the king returned in April and reassumed command of half the French forces, there had already been much progress. The French had moved ships laden with rocks and other heavy loads and sank them in a line stretching across the inlet. Eventually, enough ships were sunk so that the besieging forces were able to build a wooden dike using the sunken ships as a base. The French used thirty-seven large vessels, their prows facing out to sea, to construct the dike. To these were added fifty-nine smaller boats. They built two forts along the ends of the dike, and one large triangular wooden fort was constructed in the center. According to Baillet, this dike was the most elaborate wartime construction the French had ever undertaken, and the siege was a most impressive spectacle. This attracted many curious onlookers—young adventurers from the French nobility who wanted to observe the siege. Among them was the ever-curious René Descartes.

Descartes arrived outside the besieged city at the end of August 1628. Baillet, who is our only source on this event in the life of Descartes, tells us that Descartes came with the sole purpose of observing the military operation, as many other young men of his age were doing. He did not want to be a volunteer in the fighting or to take any part in it. He came to this battle more as an observer than he had done to any previous battle in his life. He was now the scientist—not the soldier. Descartes was especially interested in the physical properties of the great dike that Richelieu was building.

Descartes studied mathematically everything he saw during the siege. He spoke with the engineers who were building the dike, learning from them the technical details of the construction. He also met the French mathematician Desargues, who was "an expert on mechanics and was well appreciated by Cardinal Richelieu."

Besides fortifications and communications, Descartes was interested in the trajectories of cannonballs. Following his admired Galileo, Descartes wanted to learn about gravity and how it made objects fall to the ground, and he studied the curves the cannonballs were making in the air as mathematical functions.

With the huge construction project over, the harbor of La Rochelle was now completely blocked. No supplies could be brought in, and the people of the city began to starve. The fall of La Rochelle happened in stages. On September 10, the citizens sent a delegation out to meet the king at the fort in the center of the dike that was starving them out. The delegates threw themselves at the king's feet, and it seemed that an arrangement could be made. But a few days later they decided that the British fleet might still save them, and the deal was off. The French forces tightened their grip around the city, furious at the new development. In October, the British fleet indeed attempted to break through the blockade, encouraged by favorable winds. But the fleet was defeated by the French forces at the dike, and the British sought and obtained a fifteen-day cease-fire with the French. It now seemed that the city was doomed, and the two sides met to arrange a retreat of the British fleet. As British officers discussed terms of the cessation of hostilities with their French counterparts, Descartes was there, meeting the British officers. This was his first meeting with English people. He knew that there were excellent scientists in Britain and hoped to meet them. According to Baillet, two years later, in 1630, Descartes would travel to London for a short visit, and there would carry out observations of the curvature of the earth.

The condition of the people in the besieged city was now desperate. They had to resort to eating cats, dogs, and rats. Finally, having exhausted even this source of nourishment, some tried to eat the leather of belts and boots. Thirteen months after the siege began, in October 1628, the city surrendered to the French forces. Of the roughly twenty thousand inhabitants of La Rochelle before the siege began, only six thousand starving souls remained, barely alive, as the French forces stormed in. Still, some continued to fight as valiantly as they could, but they were no match for the well-fed French troops with their guns and ample ammunition.

After their last defeat, the Huguenots dispersed throughout Europe, moving to more tolerant countries. Some eventually came to the United States and founded the city of New Rochelle, New York.

Descartes entered La Rochelle with the king's forces the day the city fell to the French army, October 27, 1628. He was there to see the gloomy corteges—carriages laden with dead bodies were being pulled through the streets—and he saw the dying being given their last sacraments. This last battle, in which the French were fighting against their own starving Huguenot citizens, was an especially gory one. According to Baillet, "There has never been a worse spectacle since the fall of Jerusalem," the destruction of Jerusalem by the Romans in A.D. 70. Once he had seen enough, Descartes returned to Paris. He arrived in the French capital again on an auspicious day for him: Saint Martin's Eve, November 10.

Why would Descartes come to the site of such a dreadful siege, one in which so many people died of starvation or by the sword? Descartes was not an evil man, and neither did he harbor anti-Protestant feelings, as evidenced by the fact that he first enlisted as volunteer in the army of Prince Maurice, who led a Protestant army in 1619. But Descartes had always been attracted to military order and structure. He had been trained as a young child, starting at the age of eleven, by the Jesuits, who were run as a semimilitary Christian order. At the College of La Flèche Descartes was introduced to almost military-style discipline, order, and uniforms. It appears that the structure—both architectural and behavioral—of a militarylike place such as La Flèche appealed to him.

Additionally, Descartes always sought adventure and travel, and the military offered him both. Oddly, he could think clearly while war raged around him.

In the seventeenth century, war was waged in a well-ordered manner, similar to the way military parades are conducted today (which is perhaps the last vestige of the way armies behaved in the olden days). Soldiers on each side faced their enemy while arrayed in perfect rows, shooting at the enemy in unison. Modern warfare is completely different, and emphasizes camouflage, hiding, and an apparent *lack of order*, to prevent the enemy from exploiting any structure. Randomness is a key element in modern warfare, which stresses surprise and agility and unpredictable movements. The perfect order inherent in warfare in the 1600s seemed to attract Descartes' interest and appreciation. He is reported to have observed maneuvers and fighting, deriving from them a sense of harmony and symmetry. Descartes was interested in the trajectories of projectiles through the air; he was researching the fall of objects under the force of gravity. This research would inescapably lead him to the forbidden heliocentric theory of Copernicus.

Trajectories and military maneuvers and the construction of a unique dike to block off the entrance to La Rochelle harbor were strong motivations for Descartes to come to the coast of his ancestral Poitou to see for himself what was happening. And finally, Descartes was a man who always wanted to be at the right place at the right time. He was there when the emperor was crowned in Frankfurt, and he was in Venice to see the wedding of the city with the sea, and now he was in La Rochelle, where history (gory, to be sure) was being made. Descartes wanted to learn as much as he could about life, and the siege of La Rochelle was one such lesson.

Chapter 12

The Move to Holland and the Ghost of Galileo

DESCARTES RETURNED TO PARIS AT the end of October 1628, but soon left for Middelburg to visit Beeckman. He was returning to his mentor—the man who first got him interested in mathematics and started him on his way to pursuing a career as a mathematician and scientist. Descartes did not find his friend in Middelburg, and continued on to Dordrecht, where the two friends finally met each other at Beeckman's school.

Descartes shared with Beeckman his early attempts at unifying geometry with algebra. According to Beeckman's journal, Descartes was never able to meet anyone, in all his travels in Europe, who could understand his ideas about mathematics as well as Beeckman did. The two men discussed Descartes' novel methods in geometry and algebra, as well as music: chords and instruments and the transmission of sound. They also talked about the nature of light, and about gravity and other mysteries of physics. It seemed that the bond that united them—a love of learning and a belief in the power of mathematics—was stronger than ever. Descartes spent a few days with his friend, and then left. But they continued their regular communication through letters.

Shortly afterward, at the very end of 1628, Descartes suddenly moved to Holland. The reasons for this relocation have never been well understood. The move was permanent in the sense that Descartes would stay in Holland for the next two decades, and when he did leave that country, it would be to go to Sweden, not back to France. The move seems odd since Descartes was a Catholic and Holland was mostly Protestant. In his *Discourse on the Method*, published eight years later, Descartes said that he had moved to Holland because of a desire to distance himself from all the places at which he was known, and to live in a country in which a thriving, active population enjoyed the fruits of peace.

Holland also had more liberal printing laws than did France and other European countries, and this factor may have contributed to Descartes' decision to move, since he was hoping to publish some of his works. But Descartes' move may also have been motivated by a gnawing fear. His work had been moving in a dangerous direction—toward the Copernican theory of the universe—and Descartes worried that his discoveries in physics could be seen as contrary to the doctrines of the Catholic Church. It has been conjectured by the French scholar Gustave Cohen that the rumors, prevalent in Paris in the years 1623 to 1629, of a link between Descartes and the Rosicrucians had also contributed to his decision to leave France.

As time would prove, Descartes' move to Holland was a serious mistake. For there, he would suffer much more from Dutch theologians than he ever would have from Catholics. But perhaps his personality was changing as well. Descartes was becoming increasingly secretive.

When Descartes decided to leave France, Mersenne was disappointed and tried to dissuade him from leaving. But Descartes left despite all his efforts. He would spend the next twenty years wandering throughout Holland, carrying out a correspondence with European intellectuals—mostly through Mersenne, with whom he continued to discuss mathematical and philosophical problems. While Descartes was traveling through Holland, living for a few months or years in a given location, and then abruptly leaving for another town or village, he often did not give his true address to anyone but Mersenne. If he was staying in a town, he would dateline his letters from a neighboring town or large city. Since only Marin Mersenne knew exactly where Descartes was at any given time, all correspondence to Descartes from others had to go through the Minim friar. It seems that Descartes was hiding from someone or something.

Shortly after the move to Holland, Descartes experienced the first of the several conflicts that would come to characterize this period in his life.

It seems that Beeckman had also been experiencing a transformation through his interactions with Descartes. He developed a burning ambition to establish himself as a leading scientist, and perhaps even to compete with his brilliant friend. Early on in their relationship, on April 23, 1619, Descartes had written Beeckman saying that "if by chance something shall come out of me that would not be viewed with contempt, you may by all rights declare it as your own." But this was, perhaps, a manifestation of Descartes' extreme politeness and self-effacing nature as well as gratitude to a friend, rather than an invitation to Beeckman to claim credit for Descartes' achievements. At any rate, now Beeckman did claim such credit.

Soon after he last saw Descartes, Beeckman began his own correspondence with Marin Mersenne in Paris. Perhaps Descartes had facilitated the connection between the two men. The correspondence with Mersenne—the central figure in European mathematics and physical science in the century, the man who acted as a clearinghouse for all scientific work on the continent—was motivated by Beeckman's desire to show off his knowledge. Beeckman claimed to Mersenne, and through him to others, that it was he who had given Descartes many of his important ideas. Beeckman began to believe that he was the initiator of mathematical physics, and that Descartes was simply another person who could understand this new science, not its inventor.

Mersenne visited Beeckman in Holland, and continued on to pay a visit to Descartes. From Mersenne, Descartes found out about Beeckman's boastful claims of being the source of Descartes' knowledge—and he was deeply offended. Descartes immediately wrote to Mersenne:

I am very much obliged to you for calling to my attention the ingratitude of my friend. I think that the honor I had given him by writing to him has dazzled him and he thought that you might have a better opinion of him if he told you that he had been my master ten years ago. But he is completely mistaken, for what glory can there be in having taught a man who knows very little and freely admits it as I do? I will not mention any of this to him, since this is what you wish, but I would have much with which to make him ashamed, especially if I had his letter.

But Descartes was not assuaged by writing this letter to Mersenne. He wrote to Beeckman at the end of 1629, demanding that he return to him some of his papers, and severing all ties with his Dutch friend. In the middle of 1630, Mersenne visited Beeckman, and the latter showed him his journal, which he believed proved that he had, indeed, contributed to works by Descartes and that Descartes had not made all the discoveries he claimed without his help. When Descartes then received a letter from Beeckman telling him that he had shown his journal to Mersenne, proving his point, Descartes became even more angry than he had been until then. Descartes then wrote to Beeckman: "Now that I know that you are more interested in silly boasting than in true friendship and truth, let me tell you some things. . . . Undoubtedly, you were led to err by the politeness of our French language, when, be it in conversation, be it by letter, I have affirmed that I had learned many things from you."

But Beeckman persisted in his claims to the intellectuals of his day that it was he who had taught Descartes mathematics and physics and music theory, and that Descartes' ideas all originated in conversations with him. He wrote to Descartes, repeating these assertions. Apparently, Beeckman believed his own priority strongly enough to write it in his journal. Among other things, Beeckman wrote in the journal in Latin: "*Physici mathematici paucissimi*," or, "Rare are the physicists-mathematicians." But Descartes wrote to him, saying that "I have learned nothing from your imaginary physics, which you describe by the name mathematico-physics."

On October 17, 1630, the final rupture came. Descartes lost all patience with Beeckman and his claims on Descartes' glory. That day, he wrote him a letter that was devoid of his habitual politeness and finesse. Descartes denounced what he called Beeckman's "stupidity and ignorance." He added: "Now I recognize by all the evidence, from your last letter, that you have sinned not by malice but by insanity." Descartes insisted that he had never learned anything from Beeckman, except perhaps the smallest things about nature, such as "of ants and small worms."

Perhaps surprisingly, the two men still continued to write sporadically to each other, and they even met. But their friendship was never the same again—it lost most of its warmth and enthusiasm.

⊰⊱

In October 1629, Descartes started to work on a book on physics and metaphysics, which he called *Le Monde* ("The World"). But then in 1633, four years into this ambitious project and just when the treatise was ready to be published, Descartes heard about the trial of Galileo. Descartes' intellectual progress through life took him from pure mathematics to metaphysics, and from metaphysics to physics and cosmology. But the news about Galileo constituted a blow of unparalleled magnitude.

Starting a decade earlier, Descartes sought to apply the principles he had developed in algebra and geometry to address problems of the physical world. Descartes took the ideas of his predecessors and easily verified many of them—and disputed others—using his keen geometrical analysis. He formed a view of the universe that was squarely in agreement with the Copernican theory that the sun is the center of our solar system and the planets, including Earth, orbit the sun. Everything in Descartes' work in physics—his study of falling objects and gravity, and his observations of trajectories of cannonballs and bullets—agreed with this theory.

Descartes' physics was what we might call mathematical physics or theoretical physics. He deduced the laws of nature from first principles derived from mathematics. It was an intellectual exercise through which mathematics gave him answers about the physical world: the laws of falling bodies, the rotation of the earth, and the orbits of the planets around the sun. *Le Monde*, dedicated to Descartes' friend Mersenne in appreciation for all that he had done to promote the growth of science, mathematics, and philosophy, was his scientific description of the creation and the workings of the world—a revision of the book of Genesis in an attempt to reconcile science with religious belief. But just before the book was to be published, in November 1633, Descartes received the news about Galileo and he immediately canceled the publication of his book. He even "almost resolved to burn all my papers or at least not to let any person see them," as he later described his resolution.

History would show that Descartes was less vulnerable to the dangers that plagued Galileo than he might have thought. First, Descartes never baited the Inquisition the way Galileo had done in his writings (in which the church was represented by Simplicius, the simpleton). Second, Descartes lived in a country, Holland, that was less under the influence of Rome than was Galileo's Tuscany. And third, Descartes had very powerful allies. In 1637, Descartes' friend Mersenne would request and obtain from King Louis XIII the privilege of publishing without hindrance for "our beloved Descartes." But that would happen some years in the future, and when it did happen, it did not change Descartes' course of action. Descartes remained resolute not to publish his treatise on physics.

Descartes continued to study physics, but he refrained from publishing conclusions that could be objectionable to the church, and concentrated on a different form of physics. The shock of the trial of Galileo caused Descartes to move from theoretical physics—physics based on the use of mathematics—to experimental physics, that is, physics based on experiments in the real world, without a theoretical basis that might lead to conclusions that could enrage the powerful Inquisition.

In his first and most important published book, *Discourse on the Method*, which appeared in 1637, Descartes publicly explained his dilemma of *Le*

Monde. Right after the beginning of the fifth part of his *Discourse*, Descartes wrote the following.

> I would now need to speak about several questions that are considered controversial by the savants, with whom I do not wish to get embroiled; I think it would be better that I abstain from doing so, and that I say only in general terms what these issues are, and leave it to the wisest to determine whether it should be useful for the public to be informed about these issues more specifically. . . . I have taken notice of certain laws that God had thus established in nature, and of which he had imprinted such notions in our souls, that after enough reflection about them, we would not doubt that they are exactly observed in everything that is or is done in the world. Furthermore, considering the consequences of these laws, it seems that I have discovered many truths that are more useful and more important than the ones I had learned at first, or hoped to learn.

Descartes' *Le Monde* was published only in 1664—fourteen years after his death. The passage above refers to the material in chapter VII of *Le Monde*, in which Descartes described the three principal laws of nature governing the movements of bodies. These laws are speed, direction, and communication of movement in space. According to Descartes, these rules are founded on the immutability of divine law. Descartes believed that God created the laws of nature and *made them known to man*. He thus believed that the rotation of the earth and the revolution of the earth and the planets around the sun were direct consequences of the laws of conservation of momentum (to use modern terminology) and that they were self-evident.

But because of his worries about the Inquisition, Descartes could only hint at these ideas. He had to use coded or disguised language, and to keep most of his scientific deduction secret. Because of the fear of upsetting the church by lending theoretical support to the Copernican theory, Descartes was robbed of recognition of his role in discovering the laws of motion and of the conservation of momentum. Apparently, Descartes' genius was so profound that he was able to deduce the rotation of the earth and the motion of the planets by deriving first principles of physics, and then applying these rules to the solar system. He attributed the resulting laws of nature to divine decree. The deepest insight we can gain about Descartes' process of reasoning and physical analysis—and how it related to his religion and view of God in the universe—which were the key elements in the retracted *Le Monde*, can be gleaned from an important letter Descartes wrote to Mersenne on April 15, 1630. An excerpt follows.

> I would not allow myself in my physics to touch upon metaphysical questions, and particularly this one: that mathematical truths, which you call eternal, have been established by God and depend entirely on him, as does the rest of all creation. This, effectively, is like speaking about God as a Jupiter or Saturn and to subject them to the Styx and to the Destinies rather than say that these truths are independent of him. Do not worry at all, I beg of you, about assuring and publishing everywhere that it is God who has established these laws of nature, in the same way that a king establishes the laws of his realm. Now there is nothing at all that we are unable to comprehend, if our spirit inclines us to consider it, and all of these truths are innate in our spirit, in the same way that a king imprints his laws in the hearts of his subjects, if he has the power to do so.

Descartes seems to have been desperate to convince Mersenne the friar that there is no religious reason to avoid considering any theory about the laws of nature, since these laws have been given to the world by God and, further, have been imprinted in our consciousness. Still, he chose to retract the book he had written about this subject, rather than expose himself to the dangers that lurk from less open-minded clerics, theologians, and philosophers.

Descartes had been working toward a very ambitious goal—a universal science (in Latin: *mathesis universalis*). And he had achieved a landmark discovery. He was able to use geometry, an abstract discipline of pure mathematics, within his theory of the universe. Descartes was able to apply geometrical principles in physics, including optics and mechanics, and in his philosophy. But with what he perceived as terrifying events taking place in distant Rome just when he was ready to publish his discoveries, Descartes turned back. He withdrew the publication of *Le Monde;* he hid his papers on the subject; and then he encoded the information in new papers, destroying the originals. Descartes made the numbers embedded in his text ambiguous; he achieved this by writing his numbers in a way that made them indistinguishable from symbols, so that only he could understand what he had written. The thickness of the symbols he used was Descartes' key to deciphering whether a symbol was a number or an abstract sign.

Modern scholars who have analyzed the text of *Le Monde* have discerned in it what they called the "Fable of the World." Descartes was so concerned with hiding his true beliefs about nature that he even went as far as to disguise the world he was writing about. The world described in Descartes' withdrawn *Le Monde* is not our world. It is a mythical world that exists only in Descartes' own mind. It is *his* world. And in this fable of the world, the planets including Earth do revolve around the sun. Descartes could thus say anything he wanted about physics, biology, and the nature of light—one of the important topics of his book—without fear of ever being criticized. Hiding his physics by way of a "fable" was one more layer of protection Descartes built around himself. In the letter to Mersenne of November 25, 1630, Descartes wrote: "I like it very much, my fable of the world." So why did he have to take the extreme step of withdrawing this book from publication after he heard about the trial of Galileo in Rome?

The correspondence between Descartes and Mersenne reveals much about the development of science, about the relationship between these two men, about Descartes' state of mind, about the fate of *Le Monde*, and it perhaps even gives us a hint about the reasons for his move to Holland.

On February 1, 1634, Descartes wrote to Mersenne from Deventer:

> *Mon Révérend Père,*
>
> *While I don't have anything in particular to share with you, it has been more than two months that I have not received news from you, and I thought it best not to wait any longer before writing to you; as I've more than enough proof of the goodwill that you have so kindly chosen to bestow on me, I've had no occasion to ever doubt it, but still I feared that perhaps that feeling has cooled off, since I had not made good on my promise to send you something of my philosophy. But the knowledge I have of your virtues makes me hope that you will still have a good opinion of me, once you see that I have voluntarily and entirely canceled the treatise [that is, Le Monde], thus losing four years of my work, for the purpose of giving full obeisance to the church, since it defends the opinion about the movement of the earth. And anyway, since I have seen neither the*

pope nor the council ratifying such a defense, done only by the congregation of cardinals established for the censorship of books, I know well that now in France their authority is such that they can make of it an article of faith. I allow myself to say that the Jesuits have aided in the condemnation of Galileo, and all the books of P. Scheiner provide enough proof that they are not his friends. But otherwise the observations in that book furnish enough evidence for the movements he attributes to the sun, that I am of the belief that P. Scheiner himself is not of the opinion of Copernicus. I find this so astonishing that I do not dare publish my own sentiment. For me, I search for nothing but rest and the tranquillity of spirit, goals that cannot be achieved by those given to animosity. I wish only to instruct others, especially those who have already acquired some credit for false opinions and have some fear of loss lest the truth be revealed.

I am your very obedient and very affectionate servant,

Descartes

Descartes never even sent a copy of the withdrawn *Le Monde* to his friend Mersenne. Père Christoph Scheiner, to whose works Descartes refers, was a Jesuit astronomer who had published a treatise on sunspots. He was a good scientist and his analysis of sunspots was based on the earliest observations of this phenomenon. Galileo, however, ridiculed Scheiner's work, thus turning him into a bitter foe who then joined in the attacks against him.

Descartes appears to be jittery and fearful of the formidable forces he imagines to be arrayed against him and science in general. He seems to have a genuine, warm friendship with the friar, which he is eager to maintain. And yet, a careful reading of the letter leaves the reader with the feeling that Descartes is aware of the fact that his friend is also a member of the church. Descartes stops just short of an outright condemnation of the church, although his feelings are quite evident.

Descartes' following letter continues in the same vein, but his tone becomes harsher, and he is more blunt about the predicament in which he finds himself because of his views. We learn that Descartes' retracted treatise was probably a very interesting and valuable scientific piece of work providing support for the ideas of Galileo. The letter is datelined Deventer, end of February 1634.

> *Mon Révérend Père,*
>
> *I learn from your letter that the previous ones I had sent you have been lost, although I believe I have addressed them correctly. These letters explained at length the reasons why I have refrained from sending you my treatise, which I have no doubt you would have found legitimate, and far from blaming me or even going so far as not wanting to see me again, you would have been the first to exhort me, had I not been already myself resolute in my views. No doubt you know that Galileo had been convicted not long ago by the Inquisition, and that his opinion on the movement of the earth had been condemned as heresy. Now I will tell you that all the things I explain in my treatise, among which is also that opinion about the movement of the earth, all depend on one another, and all are based on certain evident truths. Nevertheless, I will not for the world stand up against the authority of the church. . . . I have the desire to live in peace and to continue on the road on which I have started.*

In the meantime, Descartes obtained a copy of Galileo's book and was able to read for himself the heresy that had put the great scientist in peril. Descartes' next letter to Mersenne is from Amsterdam, on August 14, 1634. He writes:

> *I begin to feel sorrow for not having any news from you. . . . Mr. Beeckman came here on Saturday evening and brought me the book of Galileo, but took it with him to Dort this morning, so I had the book in my hands for only thirty hours. I haven't had the*

opportunity to read it all, but I can see that he philosophizes well
about the movement of the earth, even though it is not enough to
persuade. . . . For what he says about a cannon fired parallel to the
horizon, I believe that you will find quite a measurable difference, if
you should perform this experiment. As for the other things, the
messenger did not allow me enough time to read the material to be
able to respond about these issues, and it is impossible for me to give
a resounding answer on any issue of physics after my having
explained all of these principles in my own treatise, which I have
resolved to suppress.

Descartes clearly believed that his own writings, published or withdrawn, contained the correct answers to the problems of physics. Perhaps this is what made him feel so close to Galileo, and thus to dread a similar or worse fate. In a letter from Leyden on June 11, 1640, he writes: "You write me about Galileo as if he were still alive, and I think that he's been dead for a long time now." Actually Galileo was alive at that time and lived for another two years, passing away on January 8, 1642. Descartes was apparently so distraught and fearful of a fate similar to Galileo's that even though he was ignorant of Galileo's fate, he assumed that the old Italian scientist must already have died—perhaps at the hands of the Inquisition. We see in these letters a portrait of an intellectual in the process of going into hiding.

Chapter 13

A Secret Affair

WHILE PREPARING HIS MASTERPIECE, the *Discourse on the Method*, having finally put the *Le Monde* debacle behind him, Descartes settled down to a reasonably quiet and happy life in Amsterdam. He lived in a building at (today's) 6 Westermarkt Street, near the West Church.

Descartes rented rooms in the building for himself and his valet from a man named Thomas Sergeant. His landlord had a pretty servant named Hélène Jans, who did Descartes' housekeeping. Hélène may have been a servant, but she was literate, since we know that in later years she wrote letters to Descartes. She had some education and a degree of culture. One autumn evening in 1634, while relaxing in a common area in the building at 6 Westermarkt Street, the two of them became lovers. According to Baillet, their daughter was conceived on Sunday, October 15, 1634. Clerselier reported in Paris that Descartes had told Chanut in 1644 that "it has been ten years now that God has removed me from that dangerous engagement." And we know that the baby girl, whom Descartes named Francine, meaning "Little France," was born on July 19, 1635. She was baptized in Deventer on August 8 in a Protestant ceremony, following the religion of her mother.

Baillet has this to say about this episode:

> The marriage of Monsieur Descartes is, for us, the most secret of the mysteries of the hidden life that he led outside of his country, far from his relatives and allies. This episode of his life may not have been one expected of a philosopher. But it was difficult for a man who for almost his entire life had been most interested in the workings of anatomy to practice rigorously the virtues of celibacy in order to conform with the laws that our saintly religion prescribes for those who live in bachelorhood.

After the publication of the *Discourse*, Descartes traveled to the Dutch coast, staying in the area around Haarlem, an isolated region of windswept sand dunes and grass. Having settled in this area, he sent for Hélène and Francine to join him. Descartes wrote to a friend telling him about his move and about his desire to bring his "niece" to visit him. This was how Descartes referred to Francine, since he was hiding the fact that he had a daughter. His letter also indicated that he wanted to bring Hélène to the house he was renting, perhaps to work as a maid for his landlady.

Descartes had intended to send Francine to France so that she could get a good education from a relative of the Descartes family, Mme. du Tronchet. But tragically, the girl died of scarlet fever on September 7, 1640, after three days of illness. Descartes was devastated by the death of his daughter. Years later, Descartes was attacked by a Dutch theologian named Gisbert Voetius for "having children outside the confines of marriage." But was Francine born outside of a marriage?

There are some clues that Descartes and Hélène had secretly married. The record of the birth of Francine in the registry of births in Holland lists her in a way that can very well be interpreted as referring to the daughter of a married couple. Descartes was very attached to both Hélène and their daughter, Francine, and may well have secretly married the mother. Since she was a servant, Descartes wanted their relationship—marriage or other—to be kept secret. But the girl was the love of his life. Still, he wandered and did not live with them for long. After the death of Francine, Descartes viewed his relationship with Hélène as a folly of youth, explaining apologetically that he was a man, that he was young at that time, and that he had never made a vow of chastity.

Three weeks after Francine died, Descartes left her mother in Amersfoort and moved to Leyden. He was brokenhearted by the loss, but kept his connection with Hélène for a period of time, writing letters to her from wherever he was at the time. Eventually, the relationship ceased to exist.

Now Descartes would set his sights on women he considered more in line with his aristocratic upbringing. He became interested in a woman who was nothing less than a princess.

At about the time of the death of Francine, Descartes also lost his older sister. He was completely devastated by these deaths, and was inconsolable. He sought solace in study. But since he was alone, save for his servants, he could not exchange ideas with anyone. But as Descartes spent more and more time talking with his servant Jean Gillot, a Huguenot, he discovered that the man had a talent for mathematics.

Descartes gave Gillot numerous problems to solve, and he performed admirably. It became clear to the master that the servant was better suited for mathematics than for his intended work. In a letter to his friend Constantijn Huygens (1596–1687), who was the secretary of the prince of Orange as well as a poet and amateur scientist, Descartes described Gillot as "my only disciple." Gillot later went on to study with Huygens himself, and then studied with other mathematicians of the day. Eventually, Jean Gillot became the official mathematician to the king of Portugal—a startling achievement for a man who began his working life as a servant, and eloquent testimony to the effects of association with Descartes.

The recent deaths made Descartes aware of his own mortality. Also, when he turned forty-seven, Descartes discovered gray hairs on his head and in his goatee. He was healthy, but the discovery of gray hair made him more concerned about the process of aging, and about death. He adopted a dog, which he named Monsieur Grat. People in the little towns and villages in which he lived in Holland would often see a lonely man walking his dog, lost in thought.

Descartes changed his diet. He now ate mostly vegetables and fruits, avoiding meat almost completely. He was living in seclusion, in small towns close to agricultural areas and country markets that sold fresh foodstuffs. He would send a servant daily to the nearest market to buy fresh eggs, milk, fruits, and vegetables. Descartes had good sense about medicine and nutrition. But he wanted more. Descartes wanted to discover a way to live to for over a century.

To achieve this goal, Descartes began to visit butcher shops. He didn't eat meat—he wanted corpses of animals. He would dissect animal cadavers, studying their anatomy in careful detail. Over the years, Descartes dissected hundreds of animals of different kinds. One of the pictures of a dissected animal Descartes drew is shown below. It is a copy of the original drawing, now lost, that Leibniz made at Clerselier's house in Paris.

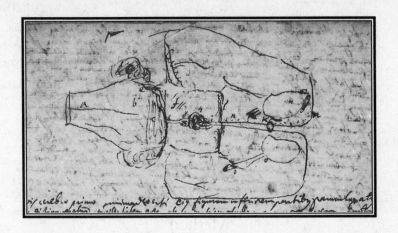

Descartes was interested in the relationship between body and soul. He dissected animals in part in order to learn anatomy and search for the secret of life in the hope of living a long time, and in part in an attempt to understand the relationship between the body and the soul. As we will see, Descartes' philosophy brought him to the belief that only people had souls.

Chapter 14

Descartes' Philosophy and the Discourse on the Method

DESCARTES IS KNOWN AS THE FA-
ther of modern philosophy thanks in large
part to the publication of his book *Discourse on the Method of Reasoning Well and
Seeking Truth in the Sciences*, in 1637. The
full title, in French, was *Discours de la
méthode pour bien conduire sa raison, et
chercher la vérité dans les sciences, plus la
Dioptrique, les Météores et la Géométrie qui
sont des essais de cette méthode*. The *Dioptrique*, dealing with Descartes' discoveries
in optics, the *Météores*, detailing his theories about natural phenomena such as the
rainbow, and the *Géométrie*, in which his
important advances in geometry and its
relation with algebra were explained, were
three appendixes demonstrating the use of
Descartes' general *method*. Descartes chose
to write the book in French to afford it the

widest possible readership among French speakers, thus following the example of Galileo, who wrote in his native Italian for a similar reason. Theirs were among the first intellectual works to be published in the vernacular, rather than the Latin of the church and the universities.

But the *Discourse* wasn't published in France. It first appeared in print on June 8, 1637, through the publisher Jean Maire in Leyden, Holland. The first printing of the book appeared anonymously.

Descartes' philosophy, which he explained in the *Discourse* (as well as in his later works), provided the basis for seventeenth-century *rationalism:* a trend toward emphasis on reason and intellect rather than emotion or imagination. Rationalism is generally contrasted with *empiricism*—the view that the main source of substantial knowledge is experience. Descartes' philosophy is based on accepting certain essential truths, not derived from experience, and seeking a system of philosophic thought based on these a priori truths while armed with a method of reasoning Descartes named "methodical doubt." Descartes viewed the mind, God, and matter as innate ideas that could not be discerned from our sensory experience in the world.

The aim of Descartes' philosophy is to use his method to reach the truth. Descartes' goal is not to discover a multiplicity of isolated truths, but rather a system of true propositions in which nothing is presupposed that is not self-evident. Thus, he insists on strong connections between all the parts of the system of knowledge he constructs. The system is thus impervious to the dangers of skepticism.

Descartes understood philosophy to mean the study of wisdom. And this wisdom meant to him a perfect knowledge of all things that human beings can know or understand. Descartes therefore included in his philosophy also metaphysics and physics and natural science. He even included in his philosophy anatomy, medicine, and morals. Descartes stressed the practical aspects of philosophy, saying that a state can have no greater good than the possession of true philosophy. Descartes deliberately broke away from the past, and was determined to start his search for truth at the beginning of all knowledge, never accepting the authority of any previous philosophy. According to Descartes, all the sciences are interconnected and must be studied as a single entity using one process designed to elicit truth. As such, his thought was at odds with the established medieval Christian philosophy, called scholasticism, which embraced Aristotelian principles and held that the various areas of knowledge are distinct from one another.

The *Discourse* was Descartes' first published book (other than the thesis he had written for his law degree in 1616). Descartes was forty-one years old when the *Discourse* appeared and became the most important, most widely read, and most controversial book of the time. People soon learned the identity of the author of this seminal work. While he had not published anything until he reached this relatively advanced age, Descartes had already written much. He had written the withdrawn *Le Monde* (whose full title was *Le Monde ou Traité de la lumière*) and had authored his *Rules (Règles pour la direction de l'esprit en la recherche de la vérité)*, which is generally believed to have been written as early as 1628. Descartes had refused to publish this work as well. The reasons for the late publication of the *Discourse*, and the withholding from publication of the *Rules*, have remained a mystery extensively debated by scholars. Here was a very private, "masked" man, reluctant to reveal to the world his deepest thoughts and theories. We know that the trial of Galileo in 1633 caused the withdrawal of *Le Monde*—but why the refusal to publish earlier, five or more years before that trial? Was fear of the Inquisition affecting Descartes even at that early period? Or were there other reasons for Descartes' behavior?

Galileo was first condemned by the church in 1616 for his support of Copernicus. And in an edict issued that same year, the Inquisition forbade the publication of any book supporting the Copernican theory in all the countries under the influence of the Catholic Church. Descartes had been aware of these developments and may already have been wary of the church and its potential response to his own work should it be published. He therefore may have decided well before Galileo's trial to refrain from letting his work become public. The news of Galileo's trial, however, greatly intensified Descartes' feeling that he had made the right decision.

The *Discourse on the Method* and its scientific appendixes reflect Descartes' painful dilemma. On the one hand, he could not allow himself to publish freely about physics, since the concepts behind his theories were all in line with those of Galileo and Copernicus, and Descartes had vowed not to contradict the views of the church. On the other hand, by 1637, Descartes felt a strong internal need to publish, and had already been under pressure from many friends and correspondents asking to read his philosophy and his views of nature.

The book and its appendixes were thus a compendium of Descartes' ideas, but one in which essential parts of physics were suppressed in order to keep the text from professing the forbidden heliocentric view. Descartes' universe, as implied by his published writings, is one that has no center and whose dimensions are infinite. These assumptions allowed Descartes to hide his true opinions and deductions about the universe and to avoid altogether the Copernican controversy. These views, however, were contrary to the scholastic tradition, according to which the universe is finite and infinity belongs to God alone.

According to recent research on Descartes' writing chronology and development of ideas, the three appendixes, the *Dioptrique*, the *Météores*, and the *Géométrie*, were formulated within the withdrawn *Le Monde*. They had thus been written some years earlier. What Descartes had done in the intervening years was to carefully reformat his work, discarding *Le Monde* and rewriting his science in such a way that it contained no traces of the forbidden physics. He then wrote a preface to the three appendixes containing his sanitized scientific writings, and published them. In fact, as evidenced from the sixth part of the *Discourse*, as well as from various letters Descartes wrote in 1633 and 1634, *Le Monde* itself was simply an *extension* of an earlier work titled *Les Météores*, dealing with a wide variety of issues in natural science, which was ready for publication in 1633. Furthermore, a treatise titled *La Dioptrique* was ready to be shipped to the printers as early as 1629, when it was withdrawn from publication. These early writings were painstakingly revised and appeared with the now-famous preface, titled the *Discourse on the Method*. Descartes' complex publishing history demonstrates the lengths to which he was ready to go in order to protect himself. His was probably the most complicated attempt to edit out controversial material in publishing history.

The *Discourse on the Method*, intended as a preface, became the main piece of writing, for it contained the principles of Descartes' philosophy. And thus this book is often published as a stand-alone treatise. The *Discourse* is unique in its format as well, as it constitutes a biographical account of the development of a philosophy—the story of the philosopher's journey of discovery.

Descartes' *Discourse on the Method* is composed of six parts. In the first part of the book, Descartes introduces his thoughts and explains how they developed. He writes about his education at the College of La Flèche and describes the ideas to which he had been exposed. "What pleased me most was mathematics, because of its certitude and its reasoning," he writes. Descartes explains how he came to believe that he could use the idea of mathematical proof in philosophy. This led him to the concept of doubt and the decision to doubt everything that he could not ascertain as true. Here the nascent Cartesian thought diverges from the accepted medieval scholastic philosophy, which held that there were three possible levels for all propositions: false, probable, and true.

In embracing purely mathematical methods of obtaining knowledge, Descartes does away with the probable and assumes the falsity of everything that he cannot prove with logical power analogous to that used in the demonstration of a theorem in geometry. He writes: "I have always had an extreme desire to learn how to distinguish the true from the false, in order to see clearly in everything that I do, and to march forward with confidence in this life." Descartes mentions his years of travel that followed his education, "thus studying from the book of the world," and concludes by stating his resolution to continue his search for truth through introspective study, never straying too far away from his books.

Descartes begins the second part of his *Discourse* by telling us that after witnessing the coronation of the new Holy Roman Emperor and joining the army in Germany, he spent the winter in an "oven," devoting his time to thinking. Among his first ideas was the notion that works by a single master are more attractive, and in a way more true to reality, than those that had been constructed by several people. From this view, he concludes that his first duty is to renounce all knowledge that had been obtained as a result of the work of many different people, that is, he wants to reject the prevailing philosophy—surely the work of many minds over generations—and to start the construction of a system of knowledge that is the work of one man alone: namely Descartes himself. All he would retain from previous knowledge would be logic, geometry, and algebra. He then states four principles that would guide him in this endeavor:

1. To accept as true only that which cannot be doubted.
2. To divide every problem into as many parts as would be necessary in order to solve it correctly.

3. To order his thoughts from the simplest to the most complicated.
4. To enumerate all concepts so that nothing pertinent is
omitted.

Descartes then discusses how problems of mathematics are solved
using his system, which is an extension of the ancient Greek method of
proving theorems by means of first principles and logical concepts. He
states his desire to be able to derive philosophical knowledge by way of
the same mathematical methodology used in geometry.

The third part of the *Discourse on the Method* is devoted to questions
of morality. Descartes tells us that he had resolved to follow the laws and
customs of the land in which he lives. He wants to be firm and resolute
in all his actions; and he wants to devote his life to cultivating his reason and rationality and to apply them in all his actions. Descartes tells
us how he returned to his travels and spent the following nine years
"rolling here and there in the world." He describes his move to Holland,
away from the places where he was known.

In the fourth part of the *Discourse*, Descartes returns to the main
thrust of his development of a philosophy. He starts with his methodical doubt: I doubt, or negate, everything that cannot be proven in a
mathematical way, he says. So what can Descartes prove? Everything is
taken to be false. But Descartes, the person, is doing this doubting. So
one thing can be deduced as true: Descartes exists. Otherwise, he could
not doubt. Thus, from the negation of everything, a proof is derived of
the existence of the person doing the doubting. This is the most brilliant
deduction in the history of Western thought. The proof is absolutely
beautiful, and it follows mathematical principles of proof. One could
even look at this deduction as a proof by contradiction—a favorite
method of mathematical proof: Assume that I do not exist. But if I don't
exist, then I could not doubt, or assume the falsity of everything in the

universe. Thus I must exist. From this deduction comes Descartes' famous "*Cogito, ergo sum*": I think, therefore I am.

The thought that I have is the primal doubt that begins the chain of deduction. I doubt everything; but this doubt is a thought; and the thought proves that I exist. I cannot doubt the fact that I am doubting; thus I, at least, must exist.

Descartes continues his logical process of deriving truth. Doubt implies uncertainty. And uncertainty implies imperfection. Human beings and everything in their environment are imperfect. But the idea of the imperfect implies the existence of something that is not imperfect. That which is not imperfect is, by definition, perfect. And perfection belongs to God. Thus Descartes deduces the existence of God from the fact that the perfect must exist. Perfect triangles and circles are geometrical figures that do not exist in our imperfect everyday world—but they do exist as ideas, as models that imperfect triangles and circles of the real world approximate. The ideal perfection implies a perfect being, God. Descartes then proceeds to the concept of a geometric space. According to Descartes, space is infinite: it extends indefinitely in all directions. Descartes' idea of a space that is unlimited in its extent leads him to the idea of infinity, and the conclusion that the infinite is God. Hence the notion that space is infinite gives Descartes another confirmation of the existence of God.

In the fifth part of the *Discourse*, Descartes turns to problems of physics and natural philosophy. He states that he cannot divulge all his beliefs about the physical world, a hint about the withdrawal of *Le Monde*. He writes about gravity, about the moon, and about the tides. His writing shows that he understands a great deal about physics. Then Descartes turns to biology and anatomy, as another example of the application of his method of reasoning. He describes the function of the heart, but incorrectly. Descartes assumes that the heart's temperature is higher than that of the rest of the body, and that the difference in temperature makes the blood flow in and out of the heart. A discussion of the function of other body parts, again flawed (he does not understand the function of the lungs, thinking their purpose is to cool the blood), leads him to the difference between animals and people. Descartes believes that language implies the existence of reason and intelligence, and hence that animals possess neither. Animals are automata, he concludes, and lack intelligence and a soul. Body and soul are separate, according to Descartes, and here again, his philosophy is at odds with scholasticism, according to which the soul is part of the body.

In the sixth and final part of the *Discourse on the Method*, Descartes states the reasons why he wrote his book. His main purpose was to contribute to the general good: he became an author in order to improve the conditions of human existence. He again returns to the dangers inherent in writing, and the fact that he could not tell us everything about his thinking and deductions about the physical world. He would accept no patronage or state pension for his work, he tells us. He wants to apply his method to the search for a deep understanding of nature with the goal of finding a way to prolong life. This aim was in keeping with the spirit of the time, the seventeenth century, when people hoped to live to the ages of the patriarchs. Finally, Descartes explains why he wrote in French rather than Latin.

But Descartes realizes that his philosophy is controversial. He is well aware of the fact that it contradicts the prevailing thought of his time. Perceptively, Descartes predicts that he would face staunch opposition for his views—but as a soldier, he is ready to defend his philosophy. He would indeed be forced to do so.

The publication of the *Discourse on the Method* made Descartes immensely famous. The book quickly elicited both positive and negative reviews by scholars, and Descartes would spend much time over the following years answering letters by many scholars about his work. His treatise became a best seller all over Europe, but the controversy resulting from this work made Descartes withdraw from people even more, interacting with the outside world mostly through letters.

In their *Fama fraternitatis* (1614), the Rosicrucians advocated "correcting the deficiencies of the church and improving moral philosophy." Descartes, who in 1637 was still unhappy about the rumors that he was a member of the Brotherhood of the Rosy Cross, continued to counter them. He wrote in his *Discourse*: "There have been as many reformers as heads," implicitly arguing against reform, to further distance himself from the Rosicrucians. He also made statements about being "suspected of this folly," elliptically alluding to the association with the brotherhood. And toward the end of the first part of his *Discourse*, Descartes puts it quite clearly:

> And finally, about bad doctrines, I thought I already knew well
> what they are worth, so as not to fall prey to deception, neither
> by the promises of an alchemist, nor by the predictions of an
> astrologer, nor by the impostures of a magician, nor by the de-
> vice of praise by those who make a profession of knowing more
> than they know.

Thus, almost two decades after he left Germany, Descartes was still concerned about the rumors connecting him with the Rosicrucians. Yet some doubts linger as one reads his writings. Earlier in the *Discourse*, Descartes writes: "I had browsed through all the books treating the topics one would consider the most *curious*" (emphasis added). According to a number of scholars, the French word *curieuses* had a particular meaning in the seventeenth century, and "curious" sciences were those dealing with special knowledge: magic, astrology, and alchemy. The words "curious science" and "curious books" appear often in the *Discourse*.

In his *Discourse*, Descartes placed hints about his secret notebook. In the second part, Descartes writes about the "analysis of the ancients and the algebra of the moderns." He alludes to the use of symbolic secret characters that are part of a "confusing and obscure art." Three pages later, he refers to an important "solution" to a problem he had found after an intense period of work. Descartes writes about a method that, applied to "arithmetic," gave him the solution, the proof, to his problem. It has been conjectured that Descartes refers here to the problem solved in his secret notebook.

Descartes' geometrical work was described in his *Géométrie*. This treatise is the most important appendix in history, for it encompassed Descartes' groundbreaking work in geometry and his wedding together of geometry and algebra—his greatest legacy in mathematics. His *Géométrie* was the "sum of all the science of pure mathematics." It would play a crucial role in the development of modern mathematics. Descartes knew that this appendix was, by far, the hardest to understand, and noted so in his text, warning the reader that an advanced level of knowledge in geometry might be required. The *Géométrie* contained extensive discussions of equations and graphs. The graphs representing the equations could not have been created without the idea of the Cartesian coordinate system, which allowed each equation to be represented with perfect precision as a curve drawn on paper. This invention was an extension of ancient Greek ideas. The Cartesian coordinate system in two dimensions is shown below.

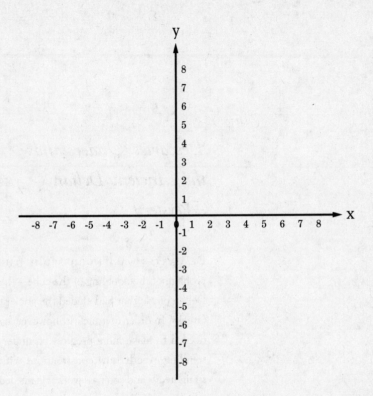

The Cartesian Coordinates

Chapter 15

Descartes Understands the Ancient Delian Mystery

DESCARTES WAS ENGROSSED IN THE problem of the doubling of the cube—the Delian puzzle that had eluded the ancient Greeks. In order to attack it, however, he needed to make more progress on understanding exactly how constructions with straightedge and compass work. He needed a tool to allow him to study these constructions, and his coordinate system constituted the required tool. Using the Cartesian coordinate system, Descartes constructed a connection between numbers and shapes—between geometry and arithmetic. The ancient Greeks had come close. For example, the Pythagoreans were able to see that the sides of a square or rectangle could be represented by numbers. This is how the Pythagorean theorem worked. If you defined the dimensions of a square as 1 unit by 1 unit, then by the Pythagorean theorem, the hypotenuse was the square root of 2. This is seen as follows.

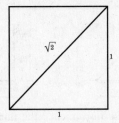

In this sense, then, one could view the distance extending to the right of the bottom left corner as 1, and the distance extending up from that corner also as 1, measured in the perpendicular direction. This gave Descartes the idea to formalize the observation of the ancient Greeks to create a *coordinate system*. Descartes realized that *any point* in the plane could now be described in terms of both its *x*-coordinate and its *y*-coordinate.

This breakthrough opened up a new world for Descartes—and would create a new world for science. But in particular, Descartes now knew what was constructible with the ancient tools of straightedge and compass, and how to construct it.

Descartes observed that a number, *a,* is constructible if we can construct two points that are a distance of *a* units apart. In his new coordinate system, a number *a* is constructible if Descartes can construct either the point $(a, 0)$ or the point $(0, a)$. Descartes observed that if *a* and *b* are constructible numbers, then so are the numbers $a+b$, $a-b$, ab, and a/b. These properties are demonstrated in the figures below.

This was a great advance. The creation of the Cartesian coordinate system immediately taught Descartes much about numbers that could be constructed with straightedge and compass. But as we will see, more

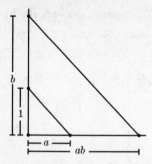
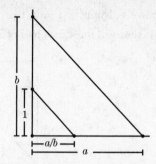

could be done with these instruments: the numbers that can be constructed are more extensive than just sums, differences, products, and quotients of constructible numbers.

Having used his new coordinate system to produce such fruitful results, Descartes took one big step forward. He was able to show that it is possible to use straightedge and compass to construct the *square root of a number*. The reason this is such an important, and perhaps unexpected, result has to do with the fields of numbers we are considering.

The rational numbers form a *field*. This means that if you start with rational numbers you stay with rational numbers: you stay within the same system, the same field, by taking sums and products and *inverses*. Every rational number—that is, either an integer or a ratio of two integers—has an inverse that is also a rational number. Simple examples are 7, whose inverse is 1/7. Or −15, whose inverse is −1/15. Or 3/19, whose inverse is 19/3. But generally, *square roots are outside the field of rational numbers* (except for the trivial cases: for example, the square root of 4 is an integer, 2). The square root of 2, for example, is outside the field of rational numbers, since no simple arithmetical operation on rational numbers (fractions whose numerators and denominators are integers) will produce the square root of 2.

But Descartes was able to show that constructions with straightedge and compass can still lead to square roots of numbers. This was one of his greatest achievements in mathematics, and the proof appears on the second page of his *Géométrie*. In providing this amazing proof, which would have stunned the ancient Greeks since they could only construct much simpler things, Descartes showed that the *field* of all the numbers that can be constructed with straightedge and compass is greater than the field of rational numbers since it now also includes square roots. Descartes was *not* able to show that this field includes cube roots of numbers, or any higher-order roots.

In fact, it would be shown two centuries after Descartes, through the work of the tormented genius Evariste Galois, a French mathematician who died in a duel at the age of twenty, that cube roots are *not* constructible, and that neither are any higher-order roots.

Descartes understood that it was this property—namely, that the straightedge and compass can go as far as square roots, but not far enough to cube roots—that made the Delian problem of the doubling of the cube impossible. It is important to note, again, that Descartes made a great stride forward by *proving* that square roots are constructible, but that he did not *prove* that cube roots are not. He "understood" that they were not, but the actual proof would require Galois Theory.

In a sense, the result could be seen as intuitive: since straightedge and compass are devices that work on the *plane*, they allow us to take *square roots* (recall that the square root of a square—a figure that lies *in the plane*—is the side of the square, also lying in the plane), but not cube roots. A *cube* naturally lives in three-dimensional space, and the *cube root* of the cube is the side.

Here is how Descartes proved that square roots are constructible with straightedge and compass: Descartes constructed, with the Greek straightedge and compass, the figure on page 166.

He now used the Pythagorean theorem *three times*. He obtained—from observing the three right triangles in the figure,

$$c^2 = a^2 + b^2$$
$$d^2 = 1^2 + b^2$$
$$(a+1)^2 = c^2 + d^2$$

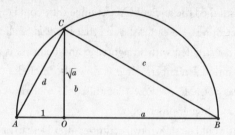

Opening up parentheses and substituting the second equation into the third, he got

$$a^2+2a+1=c^2+1^2+b^2$$

Now substituting for c^2 the sum a^2+b^2 using the first equation, he got that the equation above is

$$a^2+2a+1=a^2+b^2+1^2+b^2$$

That gave him

$$2a=2b^2$$

Or $b=\sqrt{a}$. Thus, using straightedge and compass, one can construct square roots. Voilà!

Descartes realized the following: doubling the cube was an operation in three-dimensional space, and hence the straightedge and compass— inherently *plane*, or two-dimensional, instruments—could not provide a solution. Equivalently, using algebra, he noted that to double the cube was equivalent to constructing the cube root of 2. He had proved that the square root of 2 was constructible, but he understood (although his proof was inadequate, and the correct one would come two centuries later) that the cube root was not. Descartes began to think about higher dimensions—he was captivated by the mathematical properties of the cube, and by the mystical aspects the Greeks attached to this perfect three-dimensional object.

Chapter 16

Princess Elizabeth

WHILE WORKING OUT THESE GREEK problems, Descartes continued his wanderings through Holland. He lived in Egmond, then in Santpoort, and then near Haarlem. Throughout his travels, he received mail that was forwarded to him and followed his trail. One day Descartes received a letter about a princess. Princess Elizabeth of Bohemia was also living in Holland—in exile. As a young child she escaped from Prague with her parents; as we've seen, her father, Frederick V, had been deposed as king of Bohemia right after Descartes and the victorious Bavarians and imperial troops stormed into the defeated city in 1620.

Frederick then died of the plague in Mainz in 1632, at the age of thirty-six, leaving a widow and nine children, four princesses and five princes. Princess Elizabeth was the oldest of these children. Since the deposed king's mother was the sister of Prince Maurice of Nassau (Descartes' former military employer), his widow and children had a right to seek refuge in Holland. To her dying day in exile, Elizabeth kept the title of queen of Bohemia. Her grandson would become King George I of Great Britain.

Life in exile was not easy for Elizabeth and her family. While he was still alive, Elizabeth's father, Frederick, tried from time to time to enjoy some of the pleasures he was once accustomed to as king, albeit on a much smaller scale. One day, he took his dogs and horse and went on a hunt. He was chasing a hare through the countryside when his dogs led him across a cultivated field. Before he knew it, an irate hulking peasant came out, wielding a pitchfork. "King of Bohemia, king of Bohemia!" he bellowed, apparently recognizing the exiled monarch, "you have no right to trample on my turnips like that! I'll have you know that I have worked hard to sow them." The deposed king apologized and quickly moved off the field, saying that his dogs had brought him there despite all his good intentions. J.-M. and M. Beyssade, who recount this story in *Descartes: Correspondance avec Elizabeth*, point out that in other circumstances, the rustic would have been severely punished for his insolence: in France, they surmise, he would have been put in irons; and a German prince would certainly have set his dogs on the impudent peasant.

The young Elizabeth had a thirst for knowledge, ever seeking to improve herself. She had read a Latin translation of Descartes' book the *Discourse on the Method* and wanted to learn more about his philosophy. Elizabeth was interested in all the philosophical questions that Descartes had written about. She wanted to find answers to questions about his metaphysics; she was interested in the relationship between body and soul; and she wanted to know more about his proofs of the existence of God. In addition, Elizabeth was interested in mathematics, and in particular wanted to learn how Descartes solved problems in

Greek geometry and to try her hand in solving such problems using his methods.

Princess Elizabeth was acquainted with a man whose origins were in Piedmont, named Alphonse Pollot (originally Pallotti), who had known Descartes and had renewed his friendship with the philosopher after reading Descartes' book himself. Pollot wrote to Descartes that the princess was interested in meeting him. At that time, Descartes lived not far from the deposed royals' abode. Flattered by the interest of a princess, Descartes agreed to meet her. He wrote to Pollot that he would come to her town (coincidentally named La Haye, the name of his own birthplace in France) and would "have the honor to bow to the princess and receive her orders. As for that which I hope would happen next . . ." Evidently, the aging philosopher did entertain hopes of something more.

Princess Elizabeth was twenty-four years old when she met René Descartes in 1642. At forty-six, he was almost twice her age. His nascent relationship with the princess made Descartes leave his more remote dwellings in Holland and move to Leyden and its vicinity so he could be close to her. Elizabeth became a student of Descartes' philosophy. According to Baillet, "Never had a master profited more from the penetrating mind and solidity of spirit of a disciple. Elizabeth was capable of profound meditation on the greatest mysteries of nature as well as geometry."

Elizabeth spoke perfect German like her father, perfect English like her mother, was proficient in French, and had learned Italian and Latin. She was educated in the sciences as well, and had good ability and great interest in mathematics and physics. She was described as beautiful, and looked even younger than her twenty-four years. In his letters, Descartes would describe her as an angel. She would end all her letters to the philosopher with "Your most affectionate friend to serve you."

There developed a tender relationship in which the two of them exchanged ideas. Many letters between Descartes and Elizabeth survive, and they paint a picture of a very lively and eager young woman, inter-

Princess Elizabeth

ested in learning from the older philosopher. Elizabeth was an excellent mathematician and understood Descartes' science and philosophy equally well. Descartes once said to her: "Experience has shown me that the majority of people who have the ability to understand the reasoning of metaphysics cannot conceive those of algebra; and reciprocally, that those who understand algebra are ordinarily incapable of understanding metaphysics. And I see only Your Highness as a person for whom both disciplines are equally easy to understand."

Their letters were affectionate, but they give no hint of the true nature of their relationship. The reason for the ambiguity of the letters is that Descartes and Princess Elizabeth would meet and talk face-to-face, and most of the letters between them were only written later, once she was forced to leave Holland. When that happened, her letters always passed through the hands of her siblings, and therefore intimate details could not be included.

Later events, however—having to do with Descartes' decision to move to Sweden to become tutor to Queen Christina—hint that Elizabeth may have been jealous of his new interest. Such jealousy, which was mentioned in Descartes' letters, may be an indication of a deeper attachment. Descartes guarded his privacy, making it impossible to determine the true nature of his relationship with the princess. One of his biographers, at least, did claim that Descartes and the princess had an intimate relationship. Descartes had been unable to publicly marry Hélène because she was a servant and thus socially beneath him. Equally, Princess Elizabeth was above him and most probably could not marry him. For this reason, the two kept the nature of their relationship secret. But as friends, at least, they were exceptionally close.

When Descartes moved farther north in Holland, in 1644, and was now a day's journey, rather than two hours', from La Haye, Elizabeth wrote him lamenting the distance that now separated them. But Descartes moved more and more frequently now. He felt uncomfortable staying at one location for long—perhaps experiencing the feelings of persecution from the onset of the academic controversy that would later erupt into what became known as the "Quarrel of Utrecht." And perhaps he was eager to hide from the world the true nature of his relationship with Elizabeth, and staying close to her would have betrayed the secret.

In May 1644, Descartes returned to France for an extended stay—

his first visit to his native land in sixteen years. He lodged with a friend, Abbé Picot, in the Marais—in the rue des Ecouffes between the rue du Roi de Sicilie and the rue des Francs-Bourgeois. Later, and during two other stays in Paris, Descartes rented an apartment just behind today's Place de la Contrescarpe.

Elizabeth wrote him letters to Paris, inquiring about physics and mathematics. From there, Descartes went south to the lands of his birth, visiting Blois (near Tours), Tours, Nantes, and Rennes, staying with his older brother, visiting his half-brother Joachim and his brother-in-law Roger, the widower of René's sister Jeanne. From there he wrote to the princess, promising her: "I hope, in three or four months, to have the honor of paying you a visit in La Haye."

Descartes would keep returning to visit Elizabeth, and while away from her, he would write often. But soon an invitation would come that would take him away from her, despite her protestations, and bring him to the court of a queen. There, he would take all his secrets. In his biography of Descartes, Stephen Gaukroger reports the hypothesis that Descartes left Holland for Sweden in order to intercede on Elizabeth's behalf with Queen Christina. According to this theory, Descartes was in love with Elizabeth and was heartbroken by her poverty in exile, so unfitting for a princess. Supposedly, Descartes was hoping to convince the queen of Sweden to take care of a fellow royal in distress.

Suddenly, Elizabeth had to leave Holland and seek refuge in Germany. Two of her brothers had moved to England, to stay with the royal family, their uncle and aunt. A third brother, who remained in La Haye, got involved in a personal dispute with a Frenchman from Touraine, a M. d'Espinay, who was hiding there as the result of a love scandal back in France. Elizabeth's brother and the young man got into a fight in the spice market in town, and as a result the Frenchman died. Elizabeth's mother was furious and blamed Elizabeth for inciting her brother, a charge she vehemently denied. But the mother said, "I never want to see either of you again," and the two siblings left for Germany. What was to be a relatively short stay eventually became permanent exile.

King Wladyslaw IV of Poland asked Elizabeth to marry him after his wife suddenly passed away, but she flatly rejected the royal offer. She was "in love with Descartes' philosophy," she said, and wanted to devote her life to studying it. From Berlin, Elizabeth wrote even more frequently to Descartes. These letters were exchanged through an intermediary, her younger sister Sophie. These particular letters are now lost. We do know, however, that one letter in particular contained sensitive information, for in a subsequent letter, which survives, Elizabeth asked the philosopher to burn it.

Elizabeth lived with various members of her large family, and often moved from castle to castle in Germany. For a time she lived with her brother Charles Louis, who had become an imperial elector, in his castle in Heidelberg. She also lived for a time in Brandenburg with another royal relative. She would often go with friends and relatives to Berlin to listen to music or see a play. But her greatest occupation was pursuing Descartes' philosophy—something she would continue to do after Descartes' death and her own entry into a convent in Westphalia, where she would end her days. At the convent, Elizabeth would establish a salon for Cartesian philosophy and would tell her guests that she had known the philosopher well.

Some years later, Elizabeth's sister Princess Sophie also moved to the castle in Heidelberg to live with her brother the elector. She stayed there until her marriage to the duke of Hanover, who would become Leibniz's patron. Through this connection, Sophie would develop a close friendship with Leibniz.

Chapter 17

The Intrigues of Utrecht

IN 1647, DESCARTES BECAME IN-
volved in one of the most vicious academic confrontations in history. It is hard
to understand why Descartes got into the
deep trouble he did. Great forces that had
opposed his philosophy were finally converging to attack him simultaneously in
one coordinated pile-on.

For six years, from 1641 to 1647, Descartes had been living peacefully in
the Dutch countryside, working on two
books, *Passions de l'âme* ("Passions of the
Soul," which would be published in 1649
and was concerned with the distinctions
between body and soul) and *Principes de
philosophie* ("Principles of Philosophy,"
an extension of his philosophical ideas,
published in 1647). But Descartes was increasingly being harassed by Dutch aca-

demics and others who were opposed to his philosophy on various grounds.

This philosophy became popular in the years following the publication of Descartes' *Discourse on the Method,* and Cartesianism began to be taught at universities in Europe. But since Descartes' ideas were clearly contrary to the accepted scholastic tradition—the legacy of the Middle Ages—with the interest in Cartesian principles also came a growing opposition from people who held on to the old beliefs.

Jean-Baptiste Morin (1583–1665), a mathematician, physician, and astrologer whom Descartes had befriended in Paris in the 1620s, turned against him and first attacked his work in 1638. Morin was a proponent of the geocentric theory of the universe endorsed by the church, and saw in Descartes' scientific work a dangerous form of thinking. Morin challenged Descartes' entire approach to science, questioning his results in physics. He wrote him saying that science based on mathematics must not "depend on any opinions drawn from physics." Morin sought in this way to detach science from what he saw as the evil influence of Copernican ideas in physics, fearing that they could contaminate mathematics.

Pierre Gassendi (1592–1655), a theologian and priest, was another scholar who disagreed with Descartes' writings. He questioned Descartes' philosophical work and the logical steps in Descartes' proof of existence based on his doubt, as well as his proof of the existence of God. Gassendi engaged Descartes in a series of epistolary exchanges about his philosophy. Antoine Arnauld (1612–94), another theologian and priest, raised similar objections to Descartes' philosophy. In reference to another book Descartes had authored, the *Meditations,* published in 1640, Arnauld wrote: "How does the author avoid reasoning in a circle when he says that we are sure that what we clearly and distinctly perceive is true only because God exists? We can be sure that God exists only because we clearly and distinctly perceive this." Paradoxically, despite his initial objections to Descartes' work, Arnauld went on to become an important

Cartesian philosopher. He was a prolific writer, and his work on mathematics and philosophy filled forty-three volumes.

Most of the objections to Descartes' work were raised in good spirit and led to fruitful discussions between the philosopher and his opponents. Some of them, like Arnauld, became convinced by Descartes' answers to their objections. But sometimes the opposition took the form of personal attack on Descartes himself. This was especially true in Holland, since the academics in that country knew that Descartes was living among them. Descartes did not hold an academic position himself, but his disciples did, and he was never too far from Dutch academic circles. One could view Descartes as living on the outskirts of Dutch academia. Descartes did not assume a faculty position at one of the universities because he valued his freedom to the degree that he would not have wanted to meet people on a regular basis—students and fellow faculty—as required of any professor. The University of Utrecht, however, was the closest to being the institution with which he was associated. This university taught Cartesian philosophy. Dutch theologians, however, being Protestant, often viewed Descartes' philosophical ideas as atheistic and antireligious. They tended to support scholastic and Aristotelian notions about the universe, and were therefore opposed to the new philosophy.

Descartes was a religious Catholic—and yet he was accused of atheism. This was a dangerous accusation. In 1619, a man by the name of Vanini had been burned at the stake in the French city of Toulouse for the crime of atheism. Descartes' archenemy in Holland was Gisbert Voetius (1588–1676), who perpetually accused Descartes of atheism. Ironically, Descartes had left France for Holland at least in part to escape the scrutiny of the church, and now in Holland he was being persecuted by Protestants. The Quarrel of Utrecht, as it became known, was fought through letters between Descartes and his accusers.

A key letter written by Descartes became publicly known as the Epistle to Voetius. In this letter, Descartes argued for the rights of man, and alluded to Saint Paul in 1 Corinthians 13: "All the rights of the Spirit amount to nothing without charity." This incited Voetius. A man by the name of Regius, or Henri le Roy (1598–1679), who was a disciple of Descartes and taught Cartesian philosophy at the University of Utrecht, helped Descartes in his conflict with Voetius by publishing papers supporting Descartes' ideas and by proposing Cartesian theses for public discussion.

His detractors were able to manipulate Regius, and the confrontation intensified. A friend told Descartes that his enemies behaved like pigs: "Once you grab one of them by the tail, they all squeal." And indeed, that was exactly what happened: the senate of the University of Utrecht met on March 16, 1642, and publicly condemned Cartesian philosophy, forbidding professors from teaching it at the university. All of literate Holland was engulfed in this quarrel with Descartes' philosophy. And while the man to take the blame here was ostensibly Regius, there was no doubt in anyone's mind that the proclamations were against Descartes himself—the man outside Dutch academic circles, and the creator of the new philosophy.

To make things worse for Descartes, his enemy Voetius was promoted to dean at the university and used his new powers to continue his persecution of Descartes. Since the hallmark of Descartes' philosophy is doubt, Voetius was able to manipulate it and to argue that this doubt leads to doubts about the existence of God; and hence the charge of atheism. History would judge Voetius a jealous professor who wanted to establish himself as greater than the popular Regius and chose to do so by attacking Regius's chosen philosophy, Cartesianism.

Voetius secretly authored a derogatory book about Descartes, *Admiranda methodus novae philosophiae Renati Descartes* ("An Admirable New Philosophical Method by René Descartes," published in Utrecht in 1643). The author made a key accusation against Descartes: "At first sight it is possible to say with reason that he belongs to the Society of the Brothers of the Rosy Cross." By then, Voetius had attacked the

Rosicrucians in a book published in 1639. He used his familiarity with the brotherhood to associate Descartes with Faulhaber and other purported members. And he further used the suggested association of Descartes with the Rosicrucians to strengthen his accusation of atheism against him. Within a short time, the Rosicrucians themselves would use Voetius's writings to lend legitimacy to their brotherhood by touting their imputed association with the now-famous Descartes.

The conflict evolved into a personal attack by Voetius against Descartes himself, now publicly charged as an atheist. Descartes wrote a letter to Father Dinet, a Jesuit priest who had taught at La Flèche and whom Descartes regarded highly, trying to enlist his help in the academic dispute he now faced. But in his attempts to defend himself and his philosophy, Descartes made statements that could be interpreted as attacking the moral standing of Voetius. He exposed the fact that the *Admiranda methodus* had been authored by Voetius, rather than another person, Martin Schoock, whose name appeared as the author.

Voetius retaliated in the most vicious way: he sued Descartes for defamation. The university senate and other official bodies sided with Voetius. Now the municipality of Utrecht also took action against Descartes, placing a public poster in the center of Utrecht on June 13, 1643, making public Descartes' private letters to Dinet, which they had somehow obtained, and to Voetius. Descartes was officially charged with libel against Voetius, and was left with few options. He was now in serious trouble—there was little hope of finding an honorable way out of this unexpected predicament.

But it seems that Descartes did not realize how seriously imperiled his standing was. He chose the words of the soldier that he once was, saying that while he wanted to live in peace, there was a "time to do battle." He tried to rally the forces on his side and to continue to pursue his accusers. Descartes enlisted the help of France's ambassador to Holland, but that only served to alienate him further in his adopted country. On April 10, 1644, Descartes won a minor victory in his battle. Based on new evidence provided by Descartes and his allies, the senate of the university of Utrecht cleared Descartes partially of the accusation of atheism by noting that Voetius had used false testimony from Martin Schoock in his accusations against Descartes. The latter took the new judgment by the senate and sent it to the municipality of Utrecht in an effort to clear his name. But for the municipality, the case was set and Descartes remained accused of libel against Voetius. Descartes was told by the officials that the only way he could clear his name and have the charges against him dropped was by writing an official letter of apology to Voetius.

Descartes understood that this was his only way out of the terrible conflict that could result in his being sent to prison. So on June 12, 1644, Descartes reluctantly wrote an official letter of apology to Voetius. But the conflict continued to brew because the letter, written in Latin, was not made public. Only in 1648 did the municipality of Utrecht translate Descartes' letter into French and Flemish and make it public. Descartes was left exhausted by this purposeless battle, and perhaps ready to consider moving elsewhere. This may have prepared him emotionally to accept an offer that would soon come to leave Holland for Sweden to assume the position of philosophy tutor to Queen Christina.

Interestingly, Cartesian philosophy continued to flourish in Holland in the following years. New positions in philosophy were opened at both the University of Utrecht and the University of Leyden, in which professors of Cartesian philosophy were able to teach the discipline and see it flourish. Some historians claim that Cartesian philosophy took root in Holland *despite* Descartes, rather than thanks to him. These historians note that perhaps a little more tact and diplomacy on the part of the philosopher would have gone a long way in helping him promote his teachings in Holland.

Chapter 18

The Call of the Queen

EXHAUSTED BY THE QUARREL OF
Utrecht, Descartes made another visit to
Paris, staying in the apartment at the
quiet location hidden just behind the
Place de la Contrescarpe. It was here that
Descartes made the acquaintance of
Claude Clerselier.

Like Descartes' own father, Clerselier
was a councillor to Parliament. He was
trained as an attorney but had wide-
ranging philosophical and literary interests and ambitions. Clerselier had read all
of Descartes' books and was an avid disciple of Cartesian philosophy. When he was
sixteen, Clerselier had married a twenty-
year-old woman of a wealthy family, Anne
de Virlorieux, who brought him a significant dowry and later bore him fourteen
children, many of whom died young.

Since he was well off, Clerselier could indulge his love of literature and philosophy, and he spent his days collecting books and editing and publishing works he liked and whose authors he hoped to promote. Clerselier was so obsessed with Descartes and his work that he insisted that his entire family dedicate itself to the study of Cartesian philosophy. Clerselier continually sought Descartes' attention, offering to publish and promote his works.

Soon Clerselier became Descartes' editor and translator. According to Baillet, Descartes told Clerselier "the most intimate secrets of his heart." After some time, Clerselier told Descartes that there was one member of his extended family who wanted very much to meet him—his brother-in-law Pierre Chanut. Clerselier told him that his brother-in-law had already established a great reputation for his integrity, his morals, his doctrine, and his success in business and government. All of these had led to his being regarded in the king's court as a man useful to the state.

Descartes, who in any other circumstance might have been more reluctant to meet someone new, and perhaps leery of what people's motives and intentions might have been, was favorably inclined to meet Chanut because of Clerselier's endorsement and because he had already heard of the man. Descartes' close friend Mersenne had mentioned Chanut in a letter, describing him as a person who was interested in Descartes' philosophy and had great admiration for his writings. Descartes agreed to let Clerselier arrange a meeting.

A short time after Clerselier had introduced Descartes to Chanut, the latter wrote Descartes the following.

> I write to you with such confidence that it would seem, to a
> person who doesn't know me, that an uninterrupted friendship
> of forty years, or a similarity or equality of inclinations, had
> given me this liberty. For the latter, I would vow that there is
> such a great distance from your thoughts to mine, and that I

feel I am so weak of character as compared to you, that one would be wrong to assume that you would ever love me because I resemble you in any way.

Descartes allowed himself to be thus manipulated by flattery, especially since his new acquaintance followed his letters by actions. Chanut soon became France's resident in Sweden (and later would become ambassador), and was able to lure Descartes with a significant prize: the attentions of a queen.

On November 1, 1646, Descartes wrote the following curious passage in a letter to this new friend, after Chanut had been appointed resident of France at the royal court in Stockholm:

> Mr. Clerselier has written me that you are expecting from him my *Meditations* in French in order to present them to the queen of the land in which you are. I have never had as much ambition as to desire that persons of such high rank know my name; and further, if I had only been as wise as they say the savages persuaded themselves that the monkeys were, I never would have become known as a maker of books: Since it is said that they [the savages] imagined that the monkeys could indeed speak, if they wanted to, but that they chose not to do so lest they be forced to work. And since I had not the same prudence to abstain from writing, I now have neither as much leisure nor as much peace as I would have had if I had kept quiet. But since the mistake has already been made, and since I am now known by an infinity of people at the academy, who look askance at my writings and scour them for means of harming me, I do have a great hope of being known to persons of great merit, whose power and virtue could protect me.

Apparently, Descartes was by now drained emotionally and physically from the Quarrel of Utrecht and his other persecutions in Holland and was ready to accept the protection of the queen of a far-away land.

Christina of Sweden was born on December 8, 1626, in Stockholm, the daughter of King Gustavus II Adolphus and Maria Eleonora of Brandenburg. At first, the midwives thought she was a male baby, and only as the kingdom began to celebrate the birth of the heir to the throne did people realize that in fact the baby was a little girl. Her mother was inconsolable that she had given birth to a girl, and found her daughter ugly. But the new princess, the king and queen's only child, would astonish everyone with her abilities. By the age of fifteen she had mastered Latin, French, and German, in addition to her native Swedish; eventually she would be fluent in ten languages. She read Plato and the Stoics, as well as other philosophy and literature. And she excelled in skills that were usually associated with males: equestrian sports, fencing, and archery. Christina claimed to have "an ineradicable prejudice against everything that women like to talk about or do. In women's words and occupations I showed myself to be quite incapable, and I saw no possibility of improvement in this respect."

In the Thirty Years War, France was allied with Sweden against Austria. Christina's father, Gustavus II Adolphus, led his forces into battle and was killed at the head of his cavalry. Following her father's death, Christina became the queen-elect at the age of six. Five regents headed by the prime minister, Axel Oxenstierna, governed the country. When Christina turned eighteen, in 1644, she took power. Although many significant forces opposed ending the Thirty Years War, Christina decided that it was necessary to sign the Peace of Westphalia in 1648, putting an end to it.

Christina was interested in art, music, literature, and science. She invited many leaders in these fields to come to her court. She sponsored artists and musicians, and financed hundreds of theatrical and operatic performances. For several years learned people, specialists in all branches of knowledge, arrived in Stockholm and joined her "court of learning." For this reason, Stockholm became known as "Athens of the North." After reading his *Principles of Philosophy* while on a horseback visit to an iron mine, Christina decided that she wanted Descartes to join her court of learning. Pierre Chanut, the resident of France at her court and a favored adviser, listened carefully to her as she expressed that desire, and in it he saw an opportunity for himself.

Chanut responded to Descartes' letter of November 1, 1646, saying that the queen "knows your name very well, as indeed everyone in the world should." The flattery was entirely purposeful. Bringing Descartes to Stockholm was part of a larger plan that the resident of France—soon to become the ambassador of his nation to this northern kingdom—had been hatching to increase the influence of France.

Chanut wanted to use culture to cement the alliance between France and Sweden, and Descartes fit perfectly within this scheme. The young Swedish queen was amenable to new ideas; she had a natural thirst for knowledge; and French culture held a special attraction for her. Chanut was astute enough to capitalize on her interests. He arranged for a valuable gift to be given to her by the French king: a Bible that had been specially printed for the king. She was enchanted, and Chanut told her that he knew that people in the French court were sure that she would appreciate a valuable book more than any other present.

Over the next three years, Chanut acted as intermediary in an extended mutual courtship between Christina of Sweden and René Descartes. The queen would ask Chanut to write the philosopher a letter with new questions for him, and Descartes would oblige with his answers, always addressing his letters to Chanut, but with the understanding that the information would be brought to the attention of the queen.

In December 1646, the queen asked Descartes, through Chanut, to compare bad uses of love with bad uses of hatred. Descartes obliged with a dissertation on human nature and its emotions of love and hate. Soon the queen followed with more questions. Finally came the key question: Christina, barely twenty-one years old and at the head of a nation, wanted to learn how to govern well. Through the trusted Chanut, she asked Descartes: Tell me the characteristics of a good ruler.

Descartes answered the day he received Chanut's letter. He obliged with one of his long letters back to Sweden—except that this time, he wrote directly to the queen. "I understood from M. Chanut," he wrote, "that it would please Your Majesty that I have the honor to expose to thee my opinion relevant to the Good Sovereign, considered in the sense that the ancient philosophers spoke of it." He continued by saying that God was the Good Sovereign, since he is "incomparably more perfect than his creatures." He then continued to explain over several pages the Greek views of rulers, mentioning Zeno and Epicurus. He explained to the queen his opinion that all good qualities of a ruler come from trying to emulate the attributes of God, and from trying to be close to him.

The young queen was completely taken with his answer—and with Descartes himself. After reading his response to one of her questions, she said to Chanut: "Monsieur Descartes, as much as I can see from his writing and the picture you have drawn for me, is the most fortunate of all men, and his condition is worthy of envy. You would bring me pleasure by assuring him of the great esteem I have for him."

The queen now decided that the letters were not enough. She wanted Descartes to become her private philosophy teacher. For that, he would have to leave Holland and come to her court. Chanut was elated that his plan was finally beginning to succeed. But he would have one more hurdle to pass in order to see his mission accomplished: he now needed to make Descartes want to come to Sweden.

When the queen wrote him back, Descartes was flattered. On February 26, 1649, he answered her in a letter from Egmont, Holland: "Madame, if it should have happened that a letter was sent to me from heaven, and that I saw it descend upon me from the clouds, I would not have been more surprised, and would not have received it with more respect and veneration than I did the letter Your Majesty has written me."

Descartes was still unhappy about his fights with Dutch theologians and philosophers, and in his heart still felt the hostility directed at him in his adopted country. As he wrote to Princess Elizabeth about that time, he was often going back to France and felt he was living with "one foot in each country." But he still enjoyed tranquillity in Holland, despite his troubles, and would be reluctant to leave. However, Sweden would offer him a new opportunity to start afresh, and he would enjoy being close to the seat of power there. And perhaps every philosopher wants to be able to connect his or her philosophy with worldly power: the example of Aristotle as tutor to the young Alexander the Great comes to mind. But then again, in Holland Descartes was free. He could wake up late in the morning—after sleeping his usual ten hours a night—and lazily read in bed for as long as he wanted. He lived in comfortable surroundings, usually in a small town in the countryside offering walks in nature and fresh farm food, which he valued very much, and yet never too far from a major city such as Leyden or Utrecht or Amsterdam, in which good libraries could be visited whenever he wanted, and in which intellectuals—including his many friends—could be met for discussions. This was not a life he would easily give up.

But Chanut persisted. He wrote letters to Descartes describing the queen's great intelligence, thirst for knowledge, and charm. His strongest card, however, was the queen's admiration for Descartes. Chanut wrote him: "The queen is very much concerned with your fortunes." And then, "I don't know if, once the queen tasted your philosophy, she would not try to bring you to Sweden."

Descartes responded that he was very taken with the queen's interest in him. Still, he did not want to leave Holland for what he thought was "the land of the bears, between rocks and ice," as compared with his natal "gardens of Touraine." The troubles in Holland would not be the deciding factor, since he felt he could always return to France. He was always welcome in France, and had become quite famous there—but he did not see the possibility of a major position. He described his disappointment to Elizabeth: "I believe that they want me in France as an elephant or a panther, for being a rare creature, and not at all for being useful in any way."

On February 26, 1649, Descartes responded from Holland to Chanut's letter that hinted at an upcoming formal invitation from Christina of Sweden: "Nothing ties me down to this place, other than the fact that I know nothing about another place where I might do better." Chanut continued his gentle persuasion. Finally, he wrote him most directly: "The queen of Sweden desires to see you in Stockholm and to learn philosophy from your mouth."

Reluctantly, Descartes accepted Queen Christina's invitation to come to Sweden to serve as her philosophy teacher. The queen showed great generosity and suggested, through Chanut, that Descartes take several months off before making the voyage to his new country, as well as giving him a few more months' time to get acclimated in Sweden before he would have to do anything in his new position. And with a final grand gesture, the queen ordered one of the admirals of her fleet, a man named

Flemming, to sail to Holland and pick up her royal guest and bring him to Stockholm.

When Admiral Flemming of the Royal Swedish Fleet landed in Holland in August 1649 and made the trip to Descartes' home in Egmont, the philosopher refused to follow him to his ship. Descartes claimed that he did not know who the man was and therefore would not go with him. Letters he wrote to his friends at that time provide evidence that he was still reluctant to leave Holland, and that perhaps he used this excuse in order to buy himself more time. Eventually, once letters reached him telling him that Flemming was indeed an admiral and that he had indeed been sent by the queen to pick him up, Descartes packed up his bags. He took care of his financial concerns, moving money from one bank to another, paying off debts, and revising his will. He bade farewell to his friends and was ready to leave. Some who saw him off remarked some months later that he had a premonition of his death and was ill at the thought of leaving for Sweden and an uncertain future.

On September 1, 1649, Descartes left Egmont to board his ship in Amsterdam headed for Stockholm "coiffed in curls, wearing crescent-pointed shoes, his hands covered with well-lined snow-white gloves," according to a witness. He was accompanied by his new German valet, Henry Schluter, who was fluent in French and Latin in addition to his native tongue. Before leaving, Descartes wrote to his friend Clerselier, Chanut's brother-in-law, that he was going to Sweden because of his trust in Chanut, rather than because he wanted to. And prophetically, he added: "I would be extremely upset if my presence in Stockholm should serve as the subject of malicious gossip by those who would like to say that the queen is too assiduous in her study, and that she receives her instruction from a person of *another religion*."

Descartes could not have been more correct in this concern. The intellectual court of Queen Christina had been dominated by "grammarians." These were librarians, philologists, and other scholars, all of them

Calvinist and fiercely opposed to Catholicism. They would all learn to distrust and hate Descartes, especially once he became the queen's favorite adviser.

The voyage by sea from Holland to Sweden took over a month—an unusual length of time. This was due to inclement weather accentuated by headwinds that impeded the ship's progress. According to the captain, Descartes used his science to help him navigate in difficult seas, and he said that he learned more from Descartes during the month he spent with him at sea than he had learned in decades of sailing the oceans.

Descartes arrived in Stockholm on October 4, 1649, and was received by the queen's representative. The crates containing his manuscripts and other possessions were unloaded from the ship, and he was taken to his lodgings for the night. The next day, Descartes was received by Queen Christina in a ceremony in which she lavished so much honor upon him that it incited a general jealousy of the new arrival at her court. And no one was more jealous of and hostile to the philosopher than the man named Freinsheimius, who was the queen's chief librarian.

Christina went further in honoring Descartes. She offered to grant him Swedish citizenship, and said she wanted to confer upon him noble rank. In addition, she wanted to give him lands in Germany that the Peace of Westphalia made hers. But the philosopher modestly declined these generous offers.

The queen was eager to establish the work plan for her new tutor. She wanted to meet him in the first hour after waking up in the morning—that is, at 5:00 A.M. Again, modest and polite, Descartes never told her that this was completely contrary to the lifestyle he had led all his life, staying up late at night, going to bed whenever he wanted to, never waking up before ten in the morning, and then staying in bed to read and think. So at age fifty-three, Descartes would begin a new lifestyle: one that would require him to wake up very early in the morning and leave his warm bed in the French embassy, and—in the frigid Swedish

Queen Christina of Sweden with Descartes

winter—go to the queen's unheated library, arriving there at 5:00 A.M. to deliver an hour-long lesson in philosophy. But Descartes had six weeks to get used to the idea, and to get used to living in his new country.

Descartes lived with the Chanuts, at the lodgings of the resident of France, in the embassy building. Pierre Chanut was not in Stockholm when Descartes arrived—he was in Paris for consultations with the government and also for the ceremony promoting him from resident to ambassador. Mme. Chanut, the sister of Descartes' friend Clerselier, took good care of her guest, putting him up at the upper floor of the house, located three hundred yards from the royal palace. When Pierre Chanut arrived, Descartes witnessed the ceremony in which Chanut presented his credentials as ambassador to Queen Christina. This promotion was remarkable since Chanut was a member of the middle class, and ambassadors at that time were usually members of the nobility. His promotion had much to do with merit—his success in advancing the relationship between the two nations, including the fact that he had been able to turn France's great philosopher into an adviser to the queen of Sweden.

After meeting Christina only a few times, Descartes realized that there was a strong bond between them. It was the beginning of another troubling thought for him: the possibility of a platonic, if not physical, love triangle between himself, Christina, and Elizabeth. Shortly after his arrival in Sweden, Descartes wrote the following letter to his devoted Princess Elizabeth. It was the last letter Descartes would write to her.

Stockholm, 9 October 1649

> *Madame,*
>
> *Having arrived in Stockholm four or five days ago, one of the first things I consider among my duties is to renew the offers of my very humble service to Your Highness. . . .*
>
> *One of the first things that [Queen Christina] asked me was whether I have had any news from you, and I did not fail to tell her right away that which I think about Your Highness; since, noting the strength of her spirit, I had no fear that this would give her any feelings of jealousy, as I assure myself as well that Your Highness would also not have when I openly tell you about my sentiments for this queen.*

Descartes started teaching the queen, and she proved to be the perfect student. She had endless stamina and desire to learn. While the 5:00 A.M. sessions were difficult for the philosopher, Christina could go on and on. She would spend many hours each day studying beyond her morning sessions with Descartes. While on horseback during hunts, she would carry books and read them on breaks from the chase, as well as between official duties. Christina asked Descartes about issues other than philosophy as well: literature, religion, and politics. He was becoming her chief and most favored adviser. The queen was slowly falling under the spell of Descartes, or at least that was how everyone in her court viewed the situation.

The grammarians conspired against Descartes, furious about the French and Catholic influences they saw him as exerting on their queen. Descartes felt the hostility, and it made him regret his move more than all other factors combined. On January 15, 1650, he wrote in a letter to a friend named Brégy: "The thoughts of the people here freeze during the winter as do the waters. . . . My desire to return to my desert grows every day more and more."

To another friend, he wrote that the courtiers "regard all strangers among them with much jealousy." Unfortunately, Descartes would not escape this jealousy to return to his "desert" or the gardens of Touraine.

Chapter 19

The Mysterious Death
of Descartes

ON FEBRUARY 3, 1650, FIVE MONTHS
after he arrived in Stockholm, Descartes
fell ill. Most of his biographers have concluded that his illness resulted from having to rise early—something he was not at
all used to—and from the bitter Swedish
cold. In fact, that particular winter was
the harshest in sixty years. Indeed, the
doctor who looked after Descartes in
Stockholm diagnosed his symptoms as
those of pneumonia.

Queen Christina's best doctor, her
"first doctor," was a Frenchman by nationality. His name was du Ryer, and he was a
friend of Descartes and an admirer of his
work. Du Ryer was born in Spain but had
moved to France as a young man and received his medical degree from the
University of Montpellier. He had been

introduced to Descartes' philosophy and enthusiastically called himself a Cartesian even before he met the famous philosopher at the court of Queen Christina. Descartes trusted Dr. du Ryer, and might have been inclined to follow his orders; he certainly would have been happy to have du Ryer as his physician. But the day Descartes fell ill, Dr. du Ryer was far away from Stockholm on a mission on behalf of the queen, and not expected back soon.

The queen therefore sent Descartes her "second doctor," a Dutch physician by the name of Weulles. According to Baillet, Weulles was "a sworn enemy of Descartes since the time of the war that the ministers and theologians of Utrecht and Leyden had declared upon him." Weulles had been allied with anti-Cartesian elements in Dutch academia, and was further described by Baillet as wanting to "see Descartes dead." This man "had put into use all he had judged as capable of harming" Descartes. Just why such a doctor was assigned to treat the ailing French philosopher remains a disturbing mystery.

Medical knowledge in the seventeenth century was poor, and practitioners may not have known the difference between a cold, influenza, pneumonia, or something else. To treat an illness effectively, one would need to diagnose it correctly. But whatever the illness—be it pneumonia or stomach pains or the plague—the treatment prescribed by these doctors was always the same: bleeding.

Descartes' illness started on the fifteenth day of the illness of Pierre Chanut, whom Descartes had visited every day while the ambassador was bedridden. When he left his bedside on the fifteenth day, Descartes felt a chill. That same day, Chanut had started to feel much better and was on the mend. The second day of Descartes' illness was the day of the feast of the Purification of the Virgin. Descartes was present at the ceremonies, but felt so bad that he had to leave early and get into bed.

That night, Ambassador Chanut, feeling much better himself, asked the queen to send Descartes a doctor. And according to Baillet, Weulles, "knowing the debt he had to the queen, as well as to the integrity of his profession," came to the French embassy and presented himself to Chanut to offer his services to the ailing philosopher. Descartes had avoided seeing a physician for the first two days of his illness because he was wary of charlatans and ignorant doctors. Now there was no choice—the queen had sent him Dr. Weulles, and he was too sick to refuse him.

As soon as he arrived at Descartes' bedside, Weulles decided to bleed his patient. The philosopher, who had spent time over the preceding decades studying anatomy, knew one thing that most of his contemporaries didn't: that bleeding doesn't help with anything. It can only cause infection. In fact, two years earlier, Descartes' good friend Marin Mersenne had died from an infection in his arm, which resulted from being bled as a remedy for some minor ailment.

Weulles approached Descartes, ready to cut. Descartes was surrounded by his trusted valet, Henry Schluter, as well as his friend the ambassador and the ambassador's wife. All of them urged him to allow the doctor to bleed him.

"Gentlemen," said Descartes, "spare the French blood!"

Weulles did not insist, and left the patient to his own medicine: bland food—mostly broth; water; and rest.

Descartes may well have contracted the illness from Chanut. Both men had high fever and what the doctors diagnosed as pulmonary inflammation. Chanut had been bled and was convinced that the bleeding had cured him. So after the doctor left Descartes' room, he kept pleading with him to allow Weulles to bleed him. But Descartes remained resolute against what he knew was a primitive and dangerous practice.

"Bleeding shortens our days," he said quietly, closing his eyes. Then he opened them again, and added: "I've lived forty years as an adult in good health without bleeding."

By the next day, Descartes' health had deteriorated badly. His temperature was high, and he had the worst headache he had ever experienced. His early biographers described him as feeling as though his head was about to burst. As soon as the ambassador and Mme. Chanut arrived at his bedside, they again urged the philosopher to allow the Dutch doctor to bleed him. Descartes would not hear of it. Then he said, referring to Weulles: "If I must die, I will die with more contentment if I do not see him."

Then he asked the people gathered around his bed to leave him to rest. The Chanuts and the servants all departed, leaving Descartes with only his trusted valet by his bedside.

But someone, apparently, told the Dutch doctor what Descartes had said about him, and he took it badly. It made him hate Descartes even more than he had until then. He would not cure this patient against his own will, Weulles said.

Another day passed, and the patient was still with fever and in great pain. That evening Weulles made his prognosis: the patient will die. According to Baillet, the doctor was determined to see his prediction come true.

But in the morning, Descartes felt surprisingly well. His fever had gone down and had "left his head, so his reason could return." He sat up in bed and read. He ate some bread, and he drank water. He told everyone gathered around him that he was feeling well and that the illness, apparently, had by now taken its full course. He asked to drink alcohol, requesting that it be flavored with tobacco. (Baillet surmises that this may have been Descartes' attempt to induce vomiting.)

Dr. Weulles's judgment was that this concoction would be fatal to anyone in Descartes' condition. But he said that at this stage, anything would be permitted to Descartes. So he left the room and returned with a glass filled with a dark liquid that smelled of alcohol and tobacco. He gave it to Descartes.

By the next morning, Descartes' health had taken a sharp turn for the worse. He was now throwing up blood and a blackish fluid. Phlegm came out of his mouth. He was in agony. At 8:00 A.M., Descartes, weak and about to abandon hope, finally gave up his resistance and allowed the doctor to bleed him. Very little blood came out, so an hour later, the doctor bled him again. This made the Chanuts hopeful, and they were grateful that their friend had finally agreed to the cure. But it made Descartes' condition worse. And as the day progressed, his condition deteriorated.

In the evening, while everyone else was away at dinner, Descartes asked his valet to help him out of bed so he could lie on the couch by the fire. He made it to the couch and lay there for a few moments. But he was too ill now, and the bleeding had sapped his strength. He opened his mouth and said: "Ah, my dear Schluter, this is the time I must leave."

These were Descartes' last words. He lost consciousness, and Schluter immediately rushed to get the Chanuts and the doctors and servants.

After a few more hours, it became clear that Descartes would not survive, and a priest, Father Viogué, was called in to administer the last rites. At four the next morning, Descartes died. It was the eleventh of February 1650; he was almost fifty-four years old.

Even Descartes' earliest major biographer, Adrien Baillet, mentions the rumors that began to circulate right after the philosopher's passing: that he was poisoned by Weulles in conspiracy with other members of the queen's court allied against him. (Another rumor Baillet mentions was that Descartes had died from excessive drinking of a Spanish wine, and that he wanted to die because he was heartbroken that the queen did not take well to his philosophy. Of course neither claim is likely to have been true: Descartes never drank much, and the queen was enamored with his philosophy.) Recently, a biography of Descartes by Jean-Marc Varaut, published in France in 2002, made the claim that Descartes was poisoned.

Queen Christina's court was rife with anti-Cartesian elements. Many of the people around Descartes were jealous of his place close to the queen's heart; others hated his philosophy and considered him an atheist; and there were people who despised the French philosopher because they feared his potential effect on their queen. Descartes was a Catholic and the queen and most of her subjects Lutheran. Many people feared the influence of a Catholic who was so close to the queen. And the fact that Descartes was treated by a physician who had vowed to see him dead makes the claim of a poisoning all the more believable.

And in fact, the suspicions the grammarians harbored against Descartes and his influence on their queen became a reality four years after Descartes' death: in 1654, Christina abdicated and converted to Catholicism.

Pierre Chanut, the French ambassador, who should have taken care of Descartes' interests, took inexplicable actions after the philosopher's death. The queen, who was inconsolable at the death of her dearest adviser and friend, wanted to bestow on Descartes posthumous Swedish nobility and to have him buried with the kings of Sweden. She also planned to build a grand marble mausoleum for her beloved philosopher, the man she called "My Illustrious Master."

But unexpectedly, the French ambassador opposed the idea. He argued with the queen that burying Descartes with the kings of Sweden would offend the Swedish nobility since Descartes had died a Catholic. Instead, he asked the Queen for permission to bury Descartes in the cemetery of the hospital for Orphans. Children who died before the age of reason were buried here, but also Catholics and Calvinists, since they belonged to religious minorities in the land.

The queen found the ambassador's request extremely strange, and was inclined to dismiss it. But Chanut was a persuasive man, and she was a very young, inexperienced queen. Chanut convinced her that this was for the better, since it would not be wise for the queen to antagonize her subjects, certainly not now. The ambassador knew the queen's secret—he knew that she had already been planning to abdicate and to convert to Catholicism. He used that knowledge to push her to agree to his plan. Reluctantly, the queen agreed, but insisted on paying for the expenses of Descartes' funeral. The funeral took place the next day, without much fanfare, but following the usual ceremony of a Roman Catholic burial. Descartes' coffin was borne by Chanut's eldest son and three senior officials of the French embassy. Few others were present as the philosopher's body was interred in the wretched cemetery the ambassador had insisted be Descartes' resting place.

Years after Descartes' death, on October 2, 1666, his body was exhumed, and the remains—apparently without the skull—were repatriated. They arrived in France in January 1667 and were placed in the Chapel of Saint Paul. From there, Descartes' bones were moved to a crypt in the Church of Sainte-Geneviève-du-Mont in Paris. That church was destroyed during the French Revolution. There was a move in France to rebury the remains of the great philosopher in the Panthéon with France's most distinguished citizens, and the Convention voted to approve the relocation, but the Directoire reversed that decision, and instead the body was moved to the Museum of French Monuments. Finally, in 1819, Descartes found his final resting place in the ancient Church of Saint-Germain-des-Prés.

The famous Swedish chemist Baron Jöns Jakob Berzelius (1779–1848), the discoverer of the elements cerium, selenium, and thorium, was in Paris at that time and present as Descartes' remains were reburied. He was astonished to learn that there was no skull with Descartes' bones. As fate would have it, just as Berzelius returned to Sweden, an auction was held in Stockholm in which a skull reportedly belonging to René Descartes was one of the items to be sold. Berzelius bought it.

Baron Berzelius then wrote a letter to a French baron, G. Cuvier, who was the perpetual secretary of the French Academy of Sciences. In his letter, he told Cuvier that he was donating to the French nation the skull of René Descartes, which he had just purchased, so that it "be placed with the other remains of the philosopher," clearly meaning that he wanted Descartes' skull to be buried with the rest of his bones in the Church of Saint-Germain-des-Prés. But the French baron thought otherwise. For reasons that have never been explained, once the perpetual secretary of the French Academy of Sciences received the skull, he placed it on display in a museum.

Descartes' skull, without a lower jaw and without any teeth, and with obscure writings in ink extending from the top of the cranium down to the forehead, found its ignominious end at the Musée de l'Homme (the Museum of Man) in Paris. The mangled skull reputed to be that of the great French philosopher now serves as part of a tasteless museum display about the development of the human skull. It shares its glass case with a skull marked "Cro-Magnon. Age: 100,000 years"; another, marked "Cro-Magnon. Age: 40,000 years"; a human skull of "An Early French Peasant, Homo Sapiens Sapiens. Age: 7,000 years"; and a video camera that projects the visitor's head onto a television screen, below which a caption reads: "You, Homo Sapiens Sapiens. Age: 0–120."

Descartes' skull is labeled "René Descartes, Homo Sapiens Sapiens. French Philosopher and Savant. Place of Origin: La Haye, Touraine. Emigrated to Sweden." Below it, the caption continues: "Age: 343 years (in 1993)." Under the skull and the caption is an old book, open to its title page: René Descartes, *Selected Works of Descartes*.

After Descartes was buried in Stockholm, on February 13, 1650, Pierre Chanut decided that a complete inventory of the philosopher's possessions should be made, but said that he felt that he should not be the only

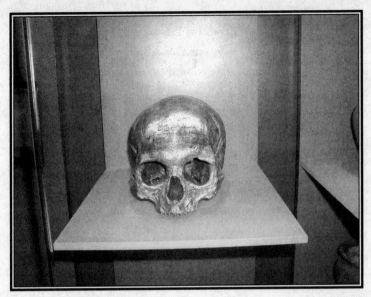

Descartes' skull

person compiling it. He therefore asked the queen to send a representative to the embassy to assist in, and be present at, the making of the inventory. The day after Descartes' funeral, February 14, the queen sent Erik Sparre, the baron of Kronoberg and president of the Court of Justice of Abo in Finland, to represent the crown at the making of the inventory. Also present were embassy officials, Father Viogué, who had been Descartes' confessor, and the deceased philosopher's valet, Henry Schluter.

All of Descartes' clothes and other personal possessions were given to Schluter, who was inconsolable at the loss of such a good master. According to Baillet, this did not prevent Schluter from making a small fortune on these items a few years later. Descartes had left a few books as well. These books were set aside to be sent to Descartes' heirs in France.

The next day these same men met again in the embassy to decide what to do with the items found in Descartes' coffer. The strongbox was opened, and found to contain several handwritten manuscripts, as well as copies of letters and documents Descartes had written and clearly deemed of importance. An inventory was then made of the materials, and Ambassador Chanut took all of these items under his "particular protection."

The items left by Descartes in his locked box shed light on what the philosopher kept close to his heart—and hidden from the world. They included the following:

> Descartes' embarrassing Letter of Apology to Voetius—a painful reminder of the debacle that helped push him out of Holland
>
> Nine volumes of copies of letters Descartes had written against Voetius
>
> Copies of Descartes' "Responses to Objections," the objections having been sent to him by his academic opponents in Holland and elsewhere
>
> Copies of letters Descartes had written to his dearest friend, Princess Elizabeth

The other items in the box were pieces of manuscripts, all handwritten and none published, with obscure titles such as *Preambles*, *Olympica*, *Democritica*, *Experimenta*, and *Parnassus*. Clearly, Descartes had kept these manuscripts for himself, never intending to have others see them. Among these manuscripts was that unique parchment notebook containing cryptic mathematical symbols, geometrical drawings, and mysterious signs that could not be identified. All the items in the inventory were labeled by Latin letters: A, B, C . . . , with the letter U always appearing with an umlaut (Charles Adam and Paul Tannery have surmised that this was Schluter's influence). The parchment notebook bearing mysterious symbols and drawings and sequences of numbers was given the designation "Item M."

The French ambassador, who did not care much about what happened to the remains of the late philosopher, allowing them to be buried in an undistinguished grave in Sweden and perhaps even failing to prevent the body from being dismembered, cared much more about the papers he had left behind. Pierre Chanut decided—without consulting with Descartes' heirs in France—to send the collection of items in the inventory as a present to his brother-in-law Claude Clerselier in Paris. But he was too busy to accomplish this task, so in the meantime the treasure would have to wait.

Two and a half years later, the ambassador prepared to leave Sweden for a new ambassadorial appointment, in Holland. He started to ship his possessions out of Sweden, and used the opportunity to send Descartes' manuscripts and letters to Clerselier. There had been pressure on the ambassador to allow the publication of some of Descartes' manuscripts, and in particular, a German biographer named Daniel Lipstorp, of Lübeck, had repeatedly asked Chanut to see Descartes' papers for use in his biography of the philosopher. But the ambassador was involved in intense diplomatic work on behalf of France in Sweden and Germany—in fact, he was even for a time in the very city of Lübeck—and had neither time nor patience for publication requests. In any case, he wanted to make sure to send everything to Clerselier; finally he made the arrangements for the shipment. The box with Descartes' hidden writings traveled to Rouen, went up the Seine to Paris, sank with the boat, and resurfaced, and the documents were saved by Clerselier.

Claude Clerselier maintained possession of all of Descartes' manuscripts until his death, in 1684, eighteen years after Leibniz had hurriedly copied parts of the manuscripts in Paris, in 1676. When Clerselier died, Descartes' papers were given to Abbé Jean-Baptiste Legrand. A short time later, the abbé responded to a request from another cleric, Père

Adrien Baillet, who asked to see the manuscripts so he could use them in the biography he was writing about Descartes. Legrand agreed to contribute as much as he could to Baillet's efforts. And it was thus that Baillet got to see Descartes' hidden manuscripts and letters. He incorporated the information in his biography of Descartes published in Paris in 1691. After over three hundred years, Baillet's biography is still the most comprehensive one ever written about Descartes.

After Legrand's death, Descartes' manuscripts disappeared. We have only the copies of some of them made by Leibniz, and a description of some of these manuscripts in Baillet's biography. But perhaps, someday, Descartes' original manuscripts may be found in the dusty and forgotten archive of some French monastery.

Baillet's biography of Descartes describes the inventory that was made in Stockholm on February 14 and 15, 1650. Adam and Tannery wrote in 1912 that a copy of the inventory was found at the French National Library in Paris. A second copy had been sent to Holland— Pierre Chanut had sent it to Descartes' friend in Holland, the scholar Constantijn Huygens. The elder Huygens wanted it for the benefit of his son, Christiaan Huygens (1629–95), who would become a famous physicist and mathematician. Through the correspondence to his father that accompanied the inventory, Christiaan Huygens would become aware that in 1653, Descartes' unpublished manuscripts had been placed in the hands of a certain Claude Clerselier living in Paris.

Descartes' death left Queen Christina distraught and deeply depressed. She became convinced that she should not stay in power. She felt empty without the comforting guidance of her philosopher—his advice on how to rule well, his ideas about the meaning of life, and the strength of his religious belief.

Within a year of Descartes' death, Queen Christina suffered a severe mental breakdown. To everyone's surprise, she frequently asked to see Father Antonio Macedo—a Catholic priest. Descartes' influence apparently extended beyond the grave and took the form of the queen's increasing attraction to his religion. In 1654, Christina abdicated and left Sweden to move to the center of Catholicism, Rome. She took a circuitous route throughout Europe, disguised as a knight, and met Catholic clergymen along the way, learning from them more about her new religious direction. She then converted to Catholicism and settled in Rome. Christina remained restless, and perhaps missed being a queen, for she attempted to seize the city of Naples and proclaim herself Queen of Naples. When the attempt failed miserably, she did not give up: she tried to secure for herself the crown of Poland. Once this attempt failed as well, Christina returned to Rome and settled there permanently. She died in 1689 and was buried in Saint Peter's Basilica.

Twelve years after the death of Descartes, the former queen of Sweden wrote:

> We certify by these presents that Sire Descartes has contributed much to our glorious conversion and that the Providence of God is served by him and by his illustrious friend Sire Chanut to give us the first lights that His grace and forgiveness would bestow on us to make us embrace the truths of the Catholic, Apostolic, and Roman religion.

The former queen made this statement public so that the world would know what had brought her to her decision. From their point of view, the grammarians were right to worry.

Chapter 20

Leibniz's Quest for Descartes' Secret

THE YEAR DESCARTES DIED IN Stockholm, 1650, a four-year-old boy in Leipzig, Germany, watched as Swedish soldiers evacuated his city, as dictated by the Peace of Westphalia, signed two years earlier, ending the Thirty Years War. France and Sweden, winning allies in the war, were to leave German soil. But as the occupation ended and the war was over, Germany began its long period of intellectual and cultural decline in the wake of the devastation brought on by the long years of fighting.

The boy, Gottfried Wilhelm Leibniz, was an unusual child. Already when he was four, those who knew him well recognized his incredible brilliance. His father, who was a professor at the University of Leipzig, knew that his son was a genius—

but unfortunately could not enjoy witnessing his son's great achievements in life because he died at age seventy, when the boy was only six years old. Gottfried read ancient Greek and Latin classics, which he found in his father's library, and within a few years had devoured books on history, art, politics, and logic. But while he was interested in many areas, he had a special ability in mathematics. And within mathematics, the young Leibniz had a very particular gift: he knew how to decipher codes.

The boy applied this unique talent both to words and to numbers. He was enamored with the mysterious, the hidden, the forbidden. His passion became the decipherment of secret messages and the quest for hidden knowledge in mathematics. Leibniz could disassemble and recombine letters to form words of a stunning number of variations and could do so with amazing speed. Similarly, he knew how to extract prime numbers by factoring numbers, and how to count and evaluate combinations. These skills lie within the area of mathematics called *combinatorics*: the study of combinations. One example of the work Leibniz did in the study of combinations was the following. He defined a notion, which he named y, as being composed of the simple notions a, b, c, and d. He then defined the following combinations: $l=ab$, $m=ac$, $n=ad$, $p=bc$, $q=bd$, and $r=cd$, as well as $s=abc$, $v=abd$, $w=acd$, and $x=bcd$. He then noted that only the combinations ax, bw, cv, ds, lr, mq, and np lead back to y. Recognizing his many gifts, Gottfried's mother sent him to an elite school, the Nicolai School in Leipzig, which he entered in 1653.

At the school, Leibniz formally learned Latin and progressed much faster than the other boys. The reason was that he had found books that were misplaced by an older student and read them voraciously. When his teachers found out that he was learning Latin outside the classroom and had already mastered the language while the other students were still struggling with the basics, they became upset and told his mother and aunts (who were helping the mother raise her son) that they should prevent the boy from reading books that were above his level. But he continued to raid his late father's library and to read more advanced books.

In 1661, after graduating from the Nicolai School at the age of fifteen, Leibniz enrolled at the University of Leipzig to study philosophy. He read the works of Aristotle, and took a course on the mathematics of Euclid from Professor Johann Kühn. This course was so difficult that Leibniz was the only student in the class who understood the subject matter. The student ended up helping the professor explain the theorems to his classmates. Leibniz also studied Bacon, Hobbes, Galileo, and Descartes.

The last of these held a special attraction for Leibniz. The young student was fascinated by Cartesian logic and philosophy, but at the same time developed his own ideas, which were often at odds with those of Descartes. According to the great English philosopher and mathematician Bertrand Russell (1872–1970), Leibniz's thinking was formed in the scholastic tradition, and he was steeped in Aristotelian-scholastic ideas about the universe. He broke away from this philosophy only through work in mathematics later in life. These beliefs may have kept Leibniz from accepting Descartes' philosophy. Additionally, Leibniz's personality was such that he could never enter any area of scholarship without being an innovator and stamping on it his own unique impression.

Leibniz had a strange mix of attraction and repulsion toward the work of Descartes, a love-hate relationship with the legacy of the late French philosopher. At that time, the opposition to Cartesian philosophy in all German universities was so high that any professor who tried to defend Cartesian ideas would be in jeopardy of losing his academic position.

Leibniz held that Descartes' principle of doubt—one of the cornerstones of Cartesian philosophy—was flawed. He wrote:

That which Descartes says about the necessity of doubting everything of which there is the least incertitude can be made more satisfying and more precise: For every concept, we must consider the degree of approval or reservation it merits. Or, more simply, one must examine the reasons for each assertion. This way we can do away with the flaws in the Cartesian doubt.

Leibniz then gives a number of examples of the flaws in Descartes' absolute doubt. If we see a combination of the colors blue and yellow, he asks in one example, can we completely doubt that the color we perceive is green? There must be a degree to such a doubt, he concludes, since when the two colors are mixed very well, the result is indeed the color green. Similarly, he asks, if one hand feels cold and the other warm, which hand should we believe? Should we completely doubt the feeling in either hand?

Leibniz wanted to learn as much as he possibly could about Descartes and was looking for more than just the published works. Years after his graduation, in 1670 and 1671, he would purchase original unpublished writings of Descartes, including letters; Descartes' *Regulae* in manuscript, purchased in Amsterdam; a manuscript entitled *Calcul de Monsieur Des Cartes*, published in 1638 as a new introduction to Descartes' *Géométrie*; and the Latin manuscript *Cartesii opera philosophica*. But he wanted more.

Leibniz wrote his master's thesis in philosophy about the relationship between philosophy and the law, and received his degree in 1664. Nine days later, his mother died. Following a graduation tinged with the sadness of losing his remaining parent, Leibniz returned to the university to study law. In 1666, he received his doctorate in law from the University of Altdorf.

Like Descartes before him, Leibniz was attracted to the work of the thirteenth-century mystic Ramon Lull. Lull's "Great Art" of combinations, which used rotating wheels within wheels to create a large number of combinations of concepts coded by letters on these wheels, took a new and deeper meaning in Leibniz's eyes. Leibniz saw in these efforts more than just mystical play, but rather a mathematical attempt to study combinations. Leibniz developed these same concepts into a mathematical theory, and published it in a 1666 treatise titled *De arte combinatoria*. This work developed the mathematical foundations of combinations, although it contained elements that had been independently discovered by Pascal in France.

Shortly afterward, in Nuremberg, Leibniz joined an alchemical society. According to Leibniz's secretary and first biographer, Johann Georg Eckhart, Leibniz wrote a letter, using obscure alchemical terminology, to the president of the alchemical society of Nuremberg. The latter was duly impressed with the writer's familiarity with alchemical secrets and offered Leibniz admittance to the society.

Leibniz began to feel that the defeated and depressed Germany of his birth could never offer him the opportunities to advance intellectually as much as other countries could, especially France—the land of high culture and advanced ideas at that time. He longed to find a way to go to Paris. His keen sense of politics would be his ticket out of Germany. Shortly after receiving his law degree, Leibniz found a patron in the distinguished German statesman Baron Johann Christian von Boineburg.

Boineburg sent Leibniz to Paris with a special mission: to try to deflect the French king, Louis XIV, from his schemes of conquering Europe. This aim was to be achieved by means of a paper Leibniz had written with Boineburg's help suggesting that France should embark on a military adventure to Egypt. It is hard to see why a French king would want to be convinced by two Germans that he should attack Egypt, but Boineburg thought that this was a possibility, and he had the money to support Leibniz's trip to Paris. Boineburg had a personal financial interest in this scheme as well: he was owed rents for his properties in France and hoped that Leibniz's mission could earn him favor with the French royal court, and that as a result he would be paid what he was owed.

Leibniz arrived in Paris at the end of March 1672, but King Louis XIV would not meet with him. Still, the young man settled down, and he met others in the French capital who had some interest in his ideas, although no progress was made on his and Boineburg's diplomatic scheme. Leibniz's own political ideas revolved around the concept of unifying Europe by reconciling its religions. He made connections with influential people and attempted to gain support for his political initiatives.

Leibniz fell in love with Paris, and over the next four years made every effort possible to make it his home. Boineburg supported him for a while, ever hopeful that Leibniz would eventually get attention in the French court. In the meantime, Leibniz was employed with tutoring Boineburg's son, a young man who was also living in Paris. But soon Boineburg died, and Leibniz needed to find other sources of income if he was to stay in Paris. Before leaving Germany, he had made contact with the duke of Hanover, whom he had sent some of his philosophical writings. He now renewed the contact, and the duke agreed to sponsor him for a while, and even gave him a letter of recommendation to help him with his endeavors. But the duke insisted that Leibniz make plans to return to Germany and serve him as librarian in Hanover. Leibniz knew that time was running out on him. There was so much he had to do, and he hoped to be able to turn down the duke's offer and stay in Paris.

Leibniz again turned his attention to mathematics. He deepened his understanding of mathematical ideas and began to develop new ones. Among these was a mechanical way of carrying out computations, which allowed him to construct a rudimentary calculating machine. Leibniz's achievements would have warranted acceptance to the French

Academy of Sciences, which would have ensured his ability to stay in the French capital. But the French declined to offer him this opportunity. They already had two foreign members in their academy: the Italian astronomer Cassini, and the Dutch physicist and mathematician Christiaan Huygens. The latter, nevertheless, became a friend of Leibniz.

Leibniz was on an intense mathematical quest. He was developing an immensely important theory, and in October 1675, he completed that theory. Leibniz indulged his fascination with Descartes by reading more of the philosopher's works. But he longed to learn ever more: he had a pressing reason to see everything he could find that had been written by Descartes.

In the spring of 1676, after he had been in Paris for over three years, Leibniz's attempts at political work collapsed completely. He no longer saw any hope of achieving success on his project of religion and diplomacy. Knowing that he would soon have no choice but to return to Germany and serve the duke of Hanover as librarian gave Leibniz's search for Descartes' manuscripts an air of urgency. He asked everyone he met whether they knew where he could find more of Descartes' writings. Finally, Christiaan Huygens told him about the inventory of Descartes' unpublished manuscripts, and gave him the name of Claude Clerselier.

So on June 1, 1676, Leibniz came to see Clerselier. He told him his story and pleaded with him to allow him to see Descartes' hidden writings. Reluctantly, the old man agreed, and Leibniz sat down and began to work.

Reading the *Preambles*, Leibniz saw Descartes' words:

Offered, once again, to the erudite scholars of the entire world,
and especially to G. F. R. C.

And in the copy he made of the manuscript, Leibniz added parenthetically the Latin name for Germany, making it read:

G. (Germania) F. R. C.

Leibniz knew very well the acronym "F. R. C." It stood for "*Fraternitas Roseae Crucis.*" Leibniz was very familiar with all the Rosicrucian writings. He knew the *Fama fraternitatis* very well, and had even discussed its finer points at length in letters. And in Leibniz's own works there is a strong Rosicrucian element that seems to be taken right out of the *Fama fraternitatis*. In 1666, in Nuremberg, Leibniz had joined the Brotherhood of the Rosy Cross. According to some sources, Leibniz was even elected secretary of the order. The order of the Rosy Cross was the parent organization of the alchemical society of Nuremberg.

We know today that the *Preambles* and *Olympica* were closely associated with the secret notebook. The contents of the lost notebook, whose title was *De solidorum elementis*, were revealed only after researchers made the decisive discovery that Leibniz had made over three centuries earlier: These private writings, which Descartes never intended to publish or share with anyone, were not separate entities. They were all pieces of a larger puzzle—the puzzle of life that Descartes wanted to solve.

Descartes' philosophy was an attempt to place human learning on a rational basis modeled on geometry. Descartes wanted people to reason in everyday life the way one reasoned in solving a mathematical problem. Within this context, the secret notebook was his crowning glory, for it contained the next level in geometry, encompassing the mystery of the universe as Descartes saw it.

Leibniz opened Descartes' secret notebook, *De solidorum elementis*, and scrutinized the writings in front of him. He didn't have much time. There were sixteen folio pages in this notebook. Either he knew that Clerselier did not want him to copy the notebook, or Clerselier had imposed very strict rules on the copying. Leibniz had to use all the mathematical skills available to him. But he had the tools needed for breaking codes—he was the expert on combinations and decoding. If anyone could break Descartes' code, it was Leibniz.

Leibniz looked at the page of Descartes' secret notebook. On one side were densely drawn figures that Descartes had sketched. It was hard to make out exactly what these were. On the other side of the page were formulas and symbols Leibniz could not immediately decipher. He quickly turned back his glance to the figures. Then he understood what

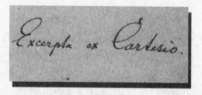

these figures were meant to represent: a cube, a pyramid, and an octahe-dron (two square-based pyramids joined at their bases).

The cube has six faces, Leibniz knew. He also knew, without having to count, since he had an immeasurably fast mind, that the pyramid has six edges. The octahedron has six vertices (or corners).

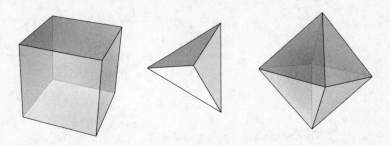

Descartes must have been on the occult search for the beast of the Apocalypse. Each of the three figures gave him a 6, and together they spelled 666. So this was Descartes' secret search—his Rosicrucian jour-ney to mystical power. Then Leibniz turned the page.

Descartes had studied the cube, a symmetric three-dimensional object whose doubling using straightedge and compass stumped the ancient Greeks. From his contacts with Faulhaber, Descartes knew that the pyramid was associated with mystical powers. He wanted to learn more about these mysterious objects. Euclid's *Elements*, translated into Latin, provided him the opportunity to do so.

Euclid wrote his *Elements* in thirteen volumes. In these volumes were the important works of Pythagoras, including his famous theorem about right triangles, work on prime numbers, and theorems about plane geometry, as well as the properties of triangles and circles. But in his thirteenth volume, Euclid devoted a major part of his writing to the *regular solids*, also called the *Platonic solids* after Plato, who knew them.

There are five Platonic solids:

1. The TETRAHEDRON, which is a pyramid with triangular faces

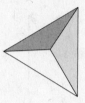

2. The CUBE, which has square faces

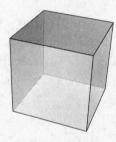

3. The OCTAHEDRON, which has triangular faces

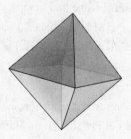

4. The DODECAHEDRON, which has pentagonal faces

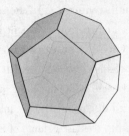

5. The ICOSAHEDRON, which has triangular faces

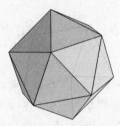

These solids are called *regular* because their faces are all the same and are congruent, and the face angles are all equal. Plato knew that there are only five such three-dimensional *polyhedra* (solids formed by plane faces). This fact made the ancient Greeks attach mystical properties to these solids, and they were believed to possess supernatural powers and to explain nature. And indeed, regular solids appear in nature. Many kinds of natural crystals are perfect (or close to perfect) regular solids. Euclid proved in volume XIII of his *Elements* many theorems about inscribing regular solids inside spheres. For example, a cube can be placed snugly inside a sphere, so that its eight vertices touch the inside surface of the sphere. The same can be done with the other regular solids. This fact proved of great importance in the late 1500s to the work of Johann Kepler.

In fact, the regular solids were known before Euclid (third century B.C.) and before Plato (fifth century B.C.). The cube, the tetrahedron, and the octahedron were known to the ancient Egyptians, whose culture predates that of the Greeks by at least a millennium. And a dodecahedron made of bronze has been found and dated to several centuries before Plato. These solids were very important in Greek mathematics, which is why sophisticated discussion and complicated theorems concerning them were placed in the last volume of Euclid's *Elements*. The regular solids were seen as the culmination of Greek geometry and its extension into three dimensions. These solids were believed to contain the secrets of the universe.

Plato visualized the five regular solids as the four elements earth, water, wind, and fire, as well as a fifth: the universe as a whole. This was a manifestation of the mystical qualities the ancient Greek mathematicians and philosophers ascribed to mathematical concepts and entities, and their view that God and the universe were mathematical. Plato visualized the five elements as in the diagram below (the fifth is the universe).

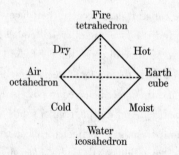

Descartes had studied the Euclidean theorems about the regular solids. But he strove to go far beyond Euclid and the ancient Greeks. The man who wedded geometry with algebra was looking for a formula that would unify all the Platonic solids, thus allowing him to extract from them a divine truth about mathematics, and perhaps about nature.

Such a crowning mathematical glory could augment and cement his philosophy.

When he turned the page, Leibniz found the remaining Platonic solids—Descartes' secret notebook contained all five of them. The number 666 was clearly not Descartes' aim. So what was it that Descartes was looking for in the Platonic solids? We know that Leibniz copied the manuscript very hastily. For a page and a half of copying, he did not see the pattern. Then suddenly he understood everything. Leibniz had found the key.

He did not need to continue the copying. All he needed to do now was to add one small marginal note—a note that none of the analysts who later read Leibniz's writing over the centuries understood, until Pierre Costabel. The mystery was now solved—there was no need to see the remaining pages of Descartes' original notebook. Leibniz now knew exactly what Descartes had discovered. And so did Pierre Costabel, the French priest and mathematician who had spent many years deciphering Leibniz's copy of Descartes' notebook, finally breaking the code in 1987.

Two decades after Leibniz copied Descartes' pages, the original notebook disappeared. On Leibniz's death in 1716, his papers were given to the archives of the Royal Library of Hanover (today's Gottfried Wilhelm Leibniz Library). Since Leibniz had left an immense amount of material, his pages copied from Descartes' notebook escaped attention for almost two centuries.

In 1860, Count Louis-Alexandre Foucher de Careil of the Sorbonne was searching through Leibniz's papers at the Hanover archives when he came upon the copy of Descartes' notebook. Foucher de Careil was not a mathematician, and he lacked the ability to decipher the secret key

Descartes had used to disguise his work. In addition, he was confounded by Descartes' bizarre notation, and mistakenly substituted for Descartes' mystical symbols, as transcribed by Leibniz, the numbers 3 and 4. This made his work even more flawed. Consequently, Foucher de Careil's report of his discovery of the notebook, published that year, was found to be useless. Foucher de Careil's work has even confused later scholars who tried to use it over the years, leading them further away from Descartes' hidden meaning. A similar fate awaited the work of two French scholars, E. Prouhet and C. Mallet, who independently attempted to decipher Descartes' secrets later that same year.

In 1890, the French Academy of Sciences prepared to republish Leibniz's pages with an explanation based on new research on the manuscript by Vice Admiral Ernest de Jonquières. But like Foucher de Careil before him, Ernest de Jonquières lacked the mathematical ability to break the secret code needed to understand Descartes' work as transcribed by Leibniz. The academy had to abandon the project.

Almost eight decades later, in 1966, new work on deciphering the notebook was done by a research group that analyzed it in conjunction with information drawn from a collection of Descartes' works compiled by Charles Adam and Paul Tannery in 1912. But yet again the notebook refused to surrender its secrets. The true meaning of the strange symbols, sequences of numbers, and unusual drawings remained an enigma.

In 1987, Pierre Costabel published his definitive analysis of Leibniz's copy of Descartes' notebook. This time, the notebook yielded its secrets. Costabel had carefully studied the notes that Leibniz had made in the margin of his copy of the notebook, and understood that Leibniz had discovered Descartes' secret key, which revealed the true meaning of Descartes' writings. The key to the mystery was finding Descartes' rule for handling the sequences of numbers in his notebook. This rule was the *gnomon*, the ancient Greek term originally meaning a staff used for casting a shadow on the ground, the length and direction of which was used to estimate time. But in Greek mathematics, the term "gnomon" came to mean a rule that specified how one was to arrange given sequences of numbers.

Descartes had analyzed the ancient Greek regular solids—Plato's mysterious objects. And within these three-dimensional geometrical objects, Descartes discovered a coveted formula: a rule that governs the structure of all of these magnificent solids. It was the Holy Grail of Greek mathematics—something the Greeks had longed to possess. But Descartes would reveal to no one this hidden truth he had discovered. Some knowledge had to be kept secret. But why keep so secret a work of geometry?

Chapter 21

Leibniz Breaks Descartes' Code and Solves the Mystery

JOHANN KEPLER KNEW THAT THE earth rotated about its axis and orbited the sun. All his astronomical work pointed in the direction of the Copernican view of the universe—the theory that Descartes espoused as well, although not in a public way. Kepler, who a few years later was to derive the laws of planetary motion—still used today in astronomy and space flights —wanted to discover the cause for the regularity he was observing in the orbits of the planets in our solar system. In doing so, while still teaching at a high school in 1595, Kepler hypothesized that there was a connection between the Greek discovery of the existence of the five regular solids and the regular orbits of the six planets known at his time (Neptune, Uranus, and Pluto were yet to be discovered).

Kepler knew from his study of Euclid's remarkable theorems in book XIII of the *Elements* that each one of the five regular solids could be perfectly inscribed inside a sphere. Motivated by his search for harmony in the creation of the solar system, he suggested the existence of celestial regular solids, whose spheres were nested. Each regular solid was inscribed in a larger sphere containing all previous regular solids and the spheres containing them. The five Platonic solids were thus placed within a sequence of nested spheres. Kepler believed that the orbits of the five other known planets (Mercury, Venus, Mars, Saturn, and Jupiter) and Earth could be viewed as circles on the surfaces of nested spheres, and he knew from Greek geometry that the five regular solids could each fit tightly inscribed inside one of these nested spheres. Kepler published this model of the solar system in his book *Mysterium cosmographicum* ("Cosmological Mystery," 1596) and considered it one of his greatest achievements, and a divine confirmation, using pure geometry, of the Copernican theory. Each sphere contained on its surface the orbit of a single planet and held inside it a regular solid. The order of these planets and solids was as follows: Mercury, octahedron, Venus, icosahedron, Earth, dodecahedron, Mars, tetrahedron, Jupiter, cube, Saturn.

The figure on page 224, from Kepler's *Mysterium cosmographicum*, shows Kepler's cosmological model using the five Platonic solids and the planets nested between them. The sun is at the center of all the spheres containing the Platonic solids and the planets.

Seeking a cosmic formula governing the Platonic solids that Kepler used to explain the universe and support the Copernican theory, Descartes began to study the mathematical properties of these ancient mystical three-dimensional geometrical objects. His purely mathematical work could thus lend theoretical support for the forbidden Copernican theory of the universe. One of the first items in Descartes' secret notebook was a theorem about placing the regular solids inside spheres, a property known to the ancient Greeks. But then Descartes went much further.

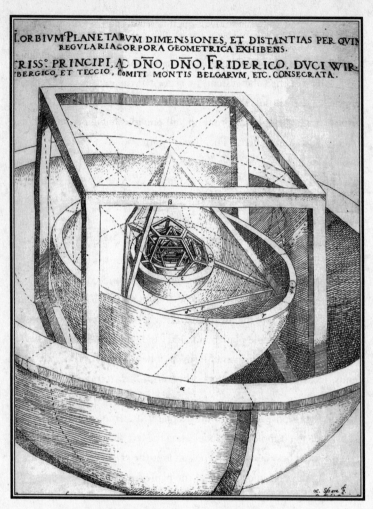

I. ORBIVM PLANETARVM DIMENSIONES, ET DISTANTIAS PER QVIN
REGVLARIA CORPORA GEOMETRICA EXHIBENS.

CRISS: PRINCIPI, AC DÑO, DÑO, FRIDERICO, DVCI WIR-
BERGICO, ET TECCIO, COMITI MONTIS BELGARVM, ETC. CONSECRATA.

Kepler's model of the universe
(*from* Mysterium cosmographicum, *1596*)

Descartes was looking for a transcendent truth that would describe all the regular solids. Later, he would discover that the formula he was after would describe not only the five regular solids but every three-dimensional polyhedron—regular or not. Descartes was interested in capturing *numerical* properties of these solids. He would then apply his overarching theory of analytic geometry to draw a link between the algebraic properties and the geometric structure of these solids. Descartes understood that the direct connection between the regular solids of ancient Greek geometry and Kepler's model of the universe could cause his work on these solids to be viewed as supporting the forbidden Copernican theory. He had to hide his work for fear of the Inquisition.

Leibniz looked at the mysterious sequences of numbers:

4   6   8   12   20   *and*   4   8   6   20   12

What was the meaning of these sequences? Leibniz saw it.

Descartes started by *counting the number of faces* of the five regular solids. He got the following sequence of numbers:

4 *(tetrahedron)*, 6 *(cube)*, 8 *(octahedron)*, 12 *(dodecahedron)*, 20 *(icosahedron)*

Then, for each one of the five regular solids, Descartes counted the number of *vertices*, or corners. This gave him, in order:

4 *(tetrahedron)*, 8 *(cube)*, 6 *(octahedron)*, 20 *(dodecahedron)*, 12 *(icosahedron)*

An inspection of the figures of the regular solids can verify these numbers. And indeed Leibniz understood that the obscure figures on the other side of the page he was looking at stood for the five regular solids.

The key to the mystery was to know what to do with the two sequences of numbers. This was Descartes' code. The key, the gnomon,

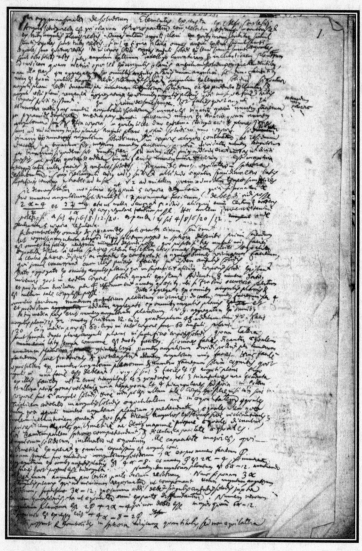

The page from Leibniz's copy of Descartes' secret notebook

told Leibniz exactly what to do with Descartes' two sequences. The rule was embedded in the way Descartes transformed and disguised other sequences of numbers in his text. Leibniz discovered the gnomon and noted it in the margin of the copy he was making. The rule told him to arrange Descartes' two sequences of numbers in an array—the second sequence below the first:

$$4 \quad 6 \quad 8 \quad 12 \quad 20$$
$$4 \quad 8 \quad 6 \quad 20 \quad 12$$

But that's where Descartes' great discovery came in. Descartes then counted the *edges* of each of the five regular solids. Let's add these as the bottom row of the array above. This gives us the following table.

	Tetrahedron	Cube	Octahedron	Dodecahedron	Icosahedron
Faces:	4	6	8	12	20
Vertices:	4	8	6	20	12
Edges:	6	12	12	30	30

Having written down the array, Descartes made his discovery. He noticed something very interesting about this array of numbers. It had to do with the sum of the first two rows as compared with the third row. (Can you see it?)

What Descartes found was that for every one of the regular solids, the sum of the number of faces and vertices minus the number of sides equals two. Or as a formula:

$$F+V-E=2$$

Then Descartes found that his formula worked for *any* three-dimensional polyhedron—regular or not. Let's check it for a square-based pyramid (not one of the five regular solids, since it has one square face and four triangular ones).

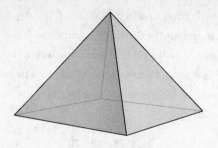

We have $F+V-E=5+5-8=2$

Descartes' formula was never attributed to him. His analysis of the three-dimensional solids would have advanced the study of geometry very much if he had published it. But because he feared the Inquisition, this important discovery remained hidden.

Descartes' formula $F+V-E=2$ is the first *topological invariant* found. The fact that the number of faces plus the number of vertices, minus the number of edges, equals 2 is a property of space itself. In deriving this formula, Descartes had thus inaugurated the immensely important mathematical field of topology. Today, topology is one of the major areas of research in mathematics, and it has important applications in physics and other fields. But since Descartes kept his discovery secret, he is credited with launching the field of analytic geometry, wedding algebra with geometry, inventing the Cartesian coordinates, and making other important advances in mathematics, but not with the founding of topology—the study of the properties of space. Others would receive the accolades for founding this field.

The Swiss mathematician Leonhard Euler (1707–83), who was born in Basel, was one of the greatest mathematicians of his century. Euler made many contributions to modern mathematics, spanning different areas within this field. Sometime after he moved to Russia to work at the Academy of Saint Petersburg, Euler discovered the magic formula that governs the structure of all three-dimensional solids: $F+V-E=2$. This equation became known as *Euler's formula*—although it might easily, as we now know, have been Descartes' formula. There is a tantalizing footnote to this story. On his way from Basel to Saint Petersburg in 1730 to assume the chair of natural sciences at the Saint Petersburg Academy, Euler passed through Hanover, Germany. He is known to have spent some time reading Leibniz's manuscripts at the Hanover archives. Whether he perused Leibniz's copy of Descartes' notebook is unknown.

What has been known as Euler's theorem and Euler's formula for over two and a half centuries is now increasingly often referred to—following the decipherment of Descartes' secret notebook by Pierre Costabel in 1987—as the Descartes-Euler theorem and the Descartes-Euler formula. But this practice is not universally followed, and many mathematicians still refer to the important property, stated as a theorem or a formula, as belonging to Euler. Had Descartes not guarded his precious knowledge so zealously, his name alone would have been attached to this discovery.

Descartes lost more than his priority in making a great mathematical discovery that founded an entire field. Despite Descartes' meticulous care during his lifetime to avoid any controversy with the church, thirteen years after his death, in 1663, his writings were placed on the *Index of Prohibited Books*. And in 1685, King Louis XIV banned the teaching of Cartesian philosophy in France. In the time of Euler, the eighteenth century, Descartes' philosophy nearly disappeared. In 1724, the Libraires Associés published the last of the older French editions of Descartes' works. For a hundred years, no new editions of Descartes' works were published in France, and the philosopher and his work were all but forgotten as new ideas emerged and philosophy was developing further. Only in 1824, exactly a century later, did his work reappear in print, and he was again recognized for his greatness as a philosopher, scientist, and mathematician. And Pierre Costabel's definitive analysis of Descartes' secret notebook a century and a half later has finally restored to Descartes his credit for founding the field of topology.

Tantalizingly, just a few decades after he died, Descartes came close to receiving the recognition he deserved for this discovery. While researching his biography of Descartes, which would appear in 1691, Baillet was trying to understand Descartes' mathematical writings, including the secret notebook, lent him by Abbé Legrand. But he could not understand any of the mysterious symbols and drawings. When he asked him, Legrand told Baillet that some years before Clerselier died, he had been visited by a young German mathematician who copied Descartes' work and may have understood the writings in the mysterious parchment notebook. Baillet then made contact with Leibniz in Hanover. Leibniz complied and explained to Baillet Descartes' mathematics. But Baillet, not being a mathematician, failed to discuss Descartes' discovery in his biography. However, he did acknowledge in his preface the help rendered him by "Monsieur Leibniz, a German mathematician."

Leibniz himself remained ambivalent in his relationship to the late French philosopher and mathematician. The man who deciphered Descartes' secret writings would grudgingly praise his work, and from some of the things he later said about Descartes, it is evident that he continually compared his abilities with those of the French genius with perhaps a degree of envy. In 1679, three years after he copied and analyzed Descartes' notebook, Leibniz wrote:

As for Descartes, this is of course not the place to praise a man the magnitude of whose genius is elevated almost above all praise. He certainly began the true and right way through ideas, and that which leads so far; but since he had aimed at his own excessive applause, he seems to have broken the thread of his investigation and to have been content with metaphysical meditations and geometrical studies by which he could draw attention to himself.

Leibniz remained obsessed with Descartes and his work for the rest of his life. He knew that Descartes had been crucial in laying down the foundations of modern science and mathematics, but he continued to argue that Descartes had stopped at a certain point in his development, believing that he, Leibniz, had gone much further. Descartes' work was clearly influencing him, and modern scholars would identify in Leibniz's philosophy both Cartesian and anti-Cartesian elements at the same time.

Leibniz remained in contact with Descartes' friends and followers. The most prominent Cartesian philosopher of the late seventeenth and early eighteenth centuries was Nicolas Malebranche (1638–1715). Malebranche first read Descartes' philosophy in a manuscript published by Clerselier in 1664. Upon reading Descartes' ideas he became so excited that he suffered from palpitations of the heart and had to be confined to bed rest. Ten years later, Malebranche wrote his own treatise on Cartesian philosophy, titled *Recherche de la vérité* ("The Search for Truth"). Leibniz exchanged letters with Princess Elizabeth, who was now sixty-one years old. He knew her through her sister Sophie, who was married to the duke of Hanover. On January 23, 1679, Leibniz wrote to Malebranche:

Through the favor of her highness, the Princess Elizabeth, who is celebrated as much for her learning as for her birth, I have been able to see [your Cartesian treatise]. . . . Descartes has said some fine things; his was a most penetrating and judicious mind. But it is impossible to do everything at once, and he has given us only some beautiful beginnings, without getting to the bottom of things. It seems to me that he is still far from the true analysis and the general art of discovery. For I am convinced that his mechanics is full of errors, that his physics goes too fast, that his geometry is too narrow, and that his metaphysics is all these things.

Why such biting and clearly unjustified criticism from the man who came so close to Descartes' monumental ideas and pursued his hidden writings? The reason was the calculus.

Before he saw Descartes' secret notebook, in the years prior to his visit to Clerselier in 1676, Leibniz was already developing his differential and integral calculus. The differential calculus is a mathematical method of finding the slope—the instantaneous rate of change—of a mathematical function. Descartes' published writings contain elements that lead in this direction. More precisely, Descartes could find the slopes of particular curves, but lacked a general, systematic method of finding slopes of functions. The integral calculus consists of an operation that is the opposite of finding slopes—integrating a mathematical function means finding the area under the curve. The ancient Greeks, especially Archimedes and Eudoxus, made advances in this area, but Leibniz found the general method.

In 1673, Leibniz traveled from Paris to London, where he met mathematicians. He so impressed British scientists and mathematicians with his work that they elected him member of the Royal Society. Eleven years later, in 1684, Leibniz published his theory of the differential calculus, followed two years later by the integral calculus. Newton, on the other hand, possessed his results on the calculus as early as 1671, although his work was not published until 1736. Leibniz's own, independent work on the calculus was completed in Paris in October 1675—before he saw any of Descartes' hidden writings.

But the calculus was not a single development—it consisted of methods and techniques that had been developed over centuries: from the ancient Greek mathematicians Archimedes and Eudoxus to Galileo, Descartes, Fermat, and others. The final glory, a unified general approach to the solution of calculus problems, was developed by Leibniz and by Newton. It therefore happened that, since Leibniz had been discussing mathematical ideas with English mathematicians as particular results were being developed, he was accused of using the ideas of others. We know today that this was not the case, and that Leibniz developed his theory of the calculus all on his own. But at the time, a controversy raged in intellectual circles in England and on the continent about the priority of related important discoveries in mathematics. For it was known even before 1736 that Isaac Newton had developed a theory of the calculus, and some had asserted that perhaps during his visit to London in 1673, Leibniz came in contact with Newton's ideas on the calculus.

Leibniz's being pressed to prove that his important discoveries were made alone and without knowledge of Newton's work also made him sensitive to any suspicion that his ideas might have been influenced by those of others—prime among them Descartes. In particular, in May 1675, English mathematicians made the claim that some of Leibniz's works in mathematics had been "nothing but deductions from Descartes." Then Leibniz received a letter in 1676 stating that "Descartes was the true founder of the new mathematical method and the contributions of his successors were only a continuation and elaboration of Descartes' ideas."

At this point, Leibniz realized that he had no choice: he had to see everything that Descartes had written—the published and the hidden, which might some day appear in print—in order to be able to defend himself and his theory of the calculus, once it was published, from any criticism whatsoever. This gave him the burning urge to search out all of Descartes' hidden works, find Clerselier, who owned these writings, and copy and understand as much as he could of all the discoveries of Descartes. He had to make sure that nothing in Descartes' writings looked too much like his own work on the calculus—or else the accusation of plagiarism could stick. It was this urgent need, and the fact that he was being attacked as having simply elaborated on the work of Descartes, that he explained to Clerselier when he came to see him in June 1676.

But the English continued to harass Leibniz with accusations of plagiarism. In August 1676, Newton wrote to Leibniz through a German interpreter, accusing him of using his work. The letter was delayed, and by the time Newton received Leibniz's answer, he was infuriated, thinking that Leibniz had taken six weeks to answer him, and hence that he was guilty. In fact, Leibniz had only a day or two to answer Newton's complaints. And he proved that his results were independent of those of Newton. He did this by showing that he had been communicated only some of Newton's specific results, and not a general method of solution. Since his calculus (and Newton's) was a very general method for the solution of a wide variety of mathematical problems, Leibniz could not have deduced it from separate, specific results that had been communicated to him by English mathematicians with whom he had ties.

Leibniz's later criticisms of Descartes' work may have been his way of further distancing himself from Descartes' genius lest he be accused of having exploited his ideas. Nothing in Descartes' work led directly to Leibniz's calculus, but Descartes' discoveries in mathematics were certainly the forerunners of the calculus.

We know that in 1661, during his first year of study at Cambridge University, Isaac Newton read books about Descartes' mathematics. Much later, after he had become a famous mathematician and scientist, Newton openly declared: "If I have seen a little farther than others, it is because I have stood on the shoulders of giants"—thus implicitly acknowledging the contributions of Galileo, Kepler, and Descartes. For without Descartes' unification of algebra and geometry it would have been impossible to describe graphs using mathematical equations, and hence, except perhaps as a pure theory, the calculus would be completely devoid of meaning.

Reluctantly, Leibniz returned to Hanover at the end of summer 1676 and spent the remaining years of his life serving the duke of Hanover in various posts. He was an educator, diplomat, counselor, and librarian. He traveled much, to Vienna, Berlin, and Italy. His final task was to write the history of the Brunswick family whom he served. When he died in 1716, that history had still not been completed. Leibniz never married. In his eulogy of Leibniz, Bernard de Fontenelle recounted that when Leibniz was fifty years old, he proposed to a lady, but that she took so much time to consider his offer that he finally withdrew it. When he died, the only heir to his considerable fortune was his sister's son. When this nephew's wife heard how much money she and her husband had inherited, she suffered a shock and died.

A Twenty-First-Century Epilogue

DESCARTES COULD BE VIEWED AS an early cosmologist—a scientist working to unveil the secrets of the universe. As such, he was the forerunner of Einstein, who in the fall of 1919—exactly three hundred years after Descartes' rapturous moment of discovery in November 1619—became the celebrity we know when his general theory of relativity was confirmed through measurements of the bending of starlight around the sun made by Arthur Eddington during a total solar eclipse a few months earlier that year. Descartes' spirit of discovery is further carried forward today by Stephen Hawking, Roger Penrose, and Alan Guth—the leading cosmologists of our own generation, expanding our horizons as we learn more about the workings of the universe.

In its essence, Descartes' work consisted of placing physics and cosmology on a firm mathematical foundation using Euclid's geometry as its base. Anyone who reads the works of modern cosmologists will be struck by the extent of the use of geometry in constructing models of the universe. The difference between the work of present-day scientists and that of Descartes is that modern cosmology is based on more advanced, specialized geometries such as the non-Euclidean geometries developed in the nineteenth century and used extensively by Einstein. Such geometries abandon Euclid's assumption that space is characterized by straight lines, and allow for a much more general structure of space, in which curves of various kinds are the basic elements.

But amazingly, the methods used by modern-day cosmologists are fundamentally extensions of those pioneered by Descartes. Physical space is so complex that in order to study its essential properties, cosmologists must rely on purely *algebraic* methods. They study the geometry of space by analyzing the properties of *groups*. A group is an abstract collection of elements with certain mathematical properties, a concept that is a direct result of the algebra studied by Descartes. And the connection between geometry and algebra, the very tool that allows modern cosmologists to carry out their advanced analysis, was, as we know, established by Descartes. But do the regular solids of ancient Greece—the elements of Descartes' most prized secret—have anything to do with cosmology?

On the eve of the transit of Venus across the face of the sun on June 8, 2004, a roughly twice-in-a-century event that was about to be observed by astronomers at the Observatory of the Aristotle University of Thessaloniki, in Greece, the American astronomer Jay M. Pasachoff delivered a lecture on the history of our understanding of the solar system.

Referring to Kepler's cosmological model based on the five Platonic solids, Pasachoff said: "It was a beautiful theoretical model of the universe. Unfortunately, it was completely wrong."

It would have seemed, therefore, that nothing about the Platonic solids had anything to do with the structure of the universe. And certainly Descartes' obsession with hiding his discovery about the nature of these solids was completely unnecessary, since Kepler's idea was not valid. The Platonic solids held no secret of the structure of the universe, and therefore did not constitute a real challenge to the earth-centered doctrine held by the church. But new research described in an article published in a mathematics journal in June 2004 may have changed everything.

On June 30, 2001, NASA launched the Wilkinson Microwave Anisotropy Probe (WMAP), a satellite designed to study minute fluctuations in the microwave background radiation that permeates all space as an echo of the Big Bang that created our universe. The fluctuations that the satellite was to study are believed to contain essential information on the geometry of the universe as a whole.

On August 10, 2001, the WMAP satellite reached its orbit far above the earth and directed its microwave antennas away from the earth into deep space. The stream of data the satellite has been producing ever since has been studied by scientists around the world.

But an enigma about the data puzzled the scientists. If the universe indeed had the infinite, flat geometry that scientists had assumed it to have, all fluctuation frequencies should have been present in the data. But surprisingly, certain fluctuations were not there. The absence of particular frequencies in the data implied to the scientists that the size of the universe was the culprit. The frequencies of the microwave background radiation that permeates space are similar in their essence to the frequencies of sound. And in the same way that the vibrations of a bell cannot be larger than the bell itself, the radiation frequencies in space are limited by the size of space itself. So cosmologists needed to look for new models of the structure of the universe: ones that would agree with the data from the satellite. These models should thus forbid the occurrence of radiation frequencies that were not found.

Complicated mathematical analysis was carried out in an effort to solve this mystery. And the answer they obtained surprised the scientists: the large-scale geometry of our universe that can answer the discrepancies in the data is a geometry based on some of the Platonic solids. It seems that while the orbits of the planets in our solar system do not follow the structure of the ancient Greek solids, models of the geometry of the entire universe do. In particular, cosmologist and MacArthur fellow Jeffrey Weeks exposed in an article in the *Notices of the American Mathematical Society* a theory that showed that tetrahedral, octahedral, and dodecahedral models of the geometry of the universe agree very well with the new findings—they completely solve the mystery of the missing fluctuations.

One new model of the geometry of the universe is thus a gigantic octahedron that is "folded onto itself" in every direction. It is an octahedron in which opposite faces are identified with each other. This means that if a spaceship travels outward from the inside of the octahedron toward one of the faces, and then passes through that face, it will arrive—speeding back *inward* into the octahedron—from exactly the opposite face of the octahedron. Another model is that of a huge icosahedron, again with opposite faces identified with each other. And a third possible model that satisfies the data is a giant dodecahedron with opposite faces identified. These models give us a universe that is closed and yet has no boundaries. Traveling (in three dimensions) within such a universe is similar to traveling on the (two-dimensional) surface of the earth: if you keep going constantly east, for example, you will travel around the world and arrive back at your starting point. You will never hit a "boundary," and you will arrive back home from "the opposite direction." Applying this principle to traveling in the "folded onto itself" dodecahedron, you will arrive back—in three-dimensional space—from the other side, that is, from the face opposite to the one you went through.

Since imagining such geometries is difficult, and since to a mathematician, one dodecahedron is exactly the same as another dodecahedron of the same size, a way of understanding the new proposed geometries of the universe is by visualizing a repeating pattern of such dodecahedrons (or octahedrons or icosahedrons). Space could thus be seen as a three-dimensional array of connected octahedrons, icosahedrons, or dodecahedrons, infinitely extended in every direction. These possible geometries of our universe are shown below.

If this theory holds under the scrutiny of other experts, and survives

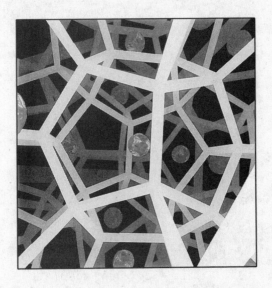

Cosmological models of the universe based on Platonic solids

the test of time, Kepler will have been proven right in assuming a connection between the Platonic solids and cosmology—albeit one he could not have envisioned. And Descartes may well have been correct in believing that the objects of his great mathematical discovery held profound cosmological relevance.

Notes

INTRODUCTION

Page

5   *"Who are we as minds in relation to our bodies?"*: Roger Ariew and Marjorie Grene, eds. *Descartes and His Contemporaries*, 1.

5–6   *which incorporated into philosophy the elements of modern psychology*: Victor Cousin, *Histoire générale de la philosophie depuis les temps les plus anciens*, 359.

7   *between the rue du Roi de Sicile and the rue des Blancs-Manteaux*: Baillet, whose 1691 biography of Descartes is still the most comprehensive after more than three hundred years, does not mention today's rue des Rosiers, which lies between the two cross streets.

PROLOGUE: LEIBNIZ'S SEARCH IN PARIS

Page

11   *found it difficult to reassemble the manuscripts*: Charles Adam and Paul Tannery, *Oeuvres de Descartes* (1974), I:xviii, after Baillet. Adam and Tannery, who quote this story from Baillet, note that Baillet would well have known exactly what happened to the papers, since he collaborated in his own biography of Descartes with Father Legrand, who received the papers from Clerselier in 1684.

14   *eagerly asked him if there was anything else*: The dates—June 1, 1676, for starting the copying and June 5, 1676, for copying the secret notebook—are men-

tioned in Henri Gouhier, *Les premières pensées de Descartes*, 14, and are based on dates inserted by Leibniz in his manuscript copy, later recopied in the nineteenth century by Foucher de Careil.

14   *imposed tight restrictions on the access to this notebook*: Pierre Costabel, ed. *René Descartes: Exercises pour les éléments des solides*, ix.

16   *Part of Leibniz's copy of Descartes' secret notebook*: Reproduced by permission from the Gottfried Wilhelm Leibniz Library in Hanover, Germany. I am indebted to Birgit Zimny for the reproduction and a copy of the entire Leibniz manuscript.

CHAPTER I: THE GARDENS OF TOURAINE
Page

18   *Joachim Descartes was the councillor of the Parliament of Brittany*: Clearly, pre-Revolution France was not a democracy, so we cannot interpret the function of a regional *Parlement* as that of, for example, the British Parliament of today. This institution had legislative and judicial roles, but they were subjected to royal authority, and these roles were more akin to those of a high court. Indeed, some of Descartes' biographers have translated the name of the institution at which Descartes' father worked as the "High Court of Rennes."

20   *there were only seventy-two Protestant baptisms in La Haye*: From information kept by the Descartes Museum in Descartes, Touraine.

21   *His baptismal certificate reads*: I am indebted to Ms. Daisy Esposito of the Descartes Museum in Descartes, Touraine, for showing me a copy of the "Acte de baptême de René Descartes" and allowing me to translate it.

21   *Her brother's wife was Jeanne (Jehanne) Proust*: This information is based on the genealogy drawn by Alfred Barbier in 1898. I am indebted to Ms. Daisy Esposito for providing me with the family tree produced by Mr. Barbier.

22   *"the land of the bears, between rocks and ice?"*: Descartes to Princess Elizabeth, April 23, 1649, quoted in Jean-Marc Varaut, *Descartes: Un cavalier français*, 256.

24   *dust rising up from the earth as it was being plowed*: Varaut, 44. See also Descartes, *Discours de la Méthode*, edited by Etienne Gilson, 108.

25   *eighteen years after René's death*: Geneviève Rodis-Lewis, *Descartes* (Paris, 1995), 18.

25   *the three-generations requirement, years after René's death*: Baillet's errors, however, are few and far between, and modern scholars have found it difficult to

disprove facts in his biography of Descartes. Generally, he seems to have done a very good job, and his biography serves as the primary source of information about the life of the philosopher- mathematician, along with the many surviving letters Descartes had written.

CHAPTER 2: JESUIT MATHEMATICS AND THE PLEASURES OF THE CAPITAL

Page

28   *right after Easter 1607*: Varaut, 48.

30   *they attended a spiritual lecture*: Varaut, 49.

30   *as well as logic, physics, and metaphysics*: Vittorio Boria, "Marin Mersenne: Educator of Scientists," 12–30.

32   *"that has thus laid them out"*: Descartes, *Discours de la méthode*, Gallimard ed., 83–84, author's translation.

34   *forever to remain in the church*: Baillet (1691), I:22.

34   *he moved to Paris*: Baillet found Descartes' year in Poitiers so uninteresting that he didn't even mention it in his biography.

36   *characterized his early days in Paris*: Baillet (1691), I:37.

38   *His friends were close to giving him up as lost*: Baillet (1691), I:36.

39   *true judgment on the evaluation of all things*: F. Alquié, ed., *Descartes: Oeuvres philosophiques*, I:46–47.

CHAPTER 3: THE DUTCH PUZZLE

Page

41   *assistant principal of the Latin School of Utrecht*: Adam and Tannery (1986), X:22.

42   *"And I suppose you will give me the solution, once you have solved this problem?"*: Adam and Tannery (1986), X:46–47.

45   *"Hence, there is no such thing as an angle"*: Adam and Tannery, (1986), X:46–47; Beeckman's *Journal*, I:237, translated in Cole, *Olympian Dreams*, 80.

46   *"at the beginning of Lent"*: Descartes to Beeckman, January 24, 1619, in F. Alquié, ed., *Descartes: Oeuvres philosophiques*, I:35.

48   *"and so on for another twenty hours"*: Descartes to Beeckman, April 29, 1619, in Alquié, *Oeuvres philosophiques*, I:42–43.

48   *new concepts could be derived*: Frances A. Yates, *The Art of Memory*, 180–84.

48   *would likely never be met*: Baillet (1692), 28.

51   *Prince Frederick of the Palatinate:* Frances A. Yates, *The Rosicrucian Enlightenment*, 23.

51   *"and are you still concerned with getting married?":* In Alquié, *Oeuvres philosophiques*, I:45.

52   *the two friends met almost every day:* Adam and Tannery (1986), X:25.

52   *"nor where I shall stop along the way":* Adam and Tannery (1986), X:162.

52   *"honor you as the promoter of this work":* Adam and Tannery (1986), X:162. This letter was found together with Beeckman's journal in 1905.

52   *Descartes was present at this magnificent ceremony:* Some modern researchers have questioned Descartes' extensive itinerary, saying it would have taken too long, given where we know he finally arrived. But Baillet is generally correct, and perhaps there is no justification for underestimating the speed at which travel could be achieved in the seventeenth century. At any rate, from Descartes' own writings in the *Discourse on the Method* (second part; Gallimard ed., 84), we know that Descartes was certainly at the coronation of the emperor.

53   *Baillet tells us:* Baillet (1691), I:63.

54   *would not carry a musket, only his sword:* Baillet (1692), 30.

CHAPTER 4: THREE DREAMS IN AN OVEN BY THE DANUBE
Page

56   *"to entertain myself with my own thoughts":* Descartes, *Discourse on the Method*, second part, in Alquié, *Oeuvres philosophiques*, I:578.

56   *used both for cooking and for heating in winter:* Varaut, 69.

58   *"an evil spirit that wanted to seduce him":* Baillet (1691), I:81.

58   *tried to take hold of the* Corpus poetarum, *it disappeared:* John R. Cole, *The Olympian Dreams and Youthful Rebellion of René Descartes*, 228 n. 14, identifies the two editions of the anthology *Corpus omnium veterum poetarum latinorum* that were available during Descartes' early years, and notes that in these editions, the two poems *"Quod vitae sectabor iter?"* and *"Est et Non"* appear either on the same page or on facing pages. It seems that Descartes' memory of reading these poems served him even in his dream.

59   *"truth and falsehood in the secular sciences":* Baillet (1691), I;82.

61   *"Anno 1619 Kalendis Januarii":* Adam and Tannery (1986), X:7.

61   *who discovered the laws of planetary motion:* Edouard Mehl, Descartes en Allemagne, 17.

61   *has conjectured that such a meeting indeed took place:* L. Gäbe, "Cartelius oder

Notes

Cartesius: Eine Korrectur zu meinem Buche über Descartes Selbstkritik, Hamburg, 1972," *Archiv für Geschichte der Philosophie* 58 (1976), 58–59.

62   *"truly worthy of your consideration"*: Quoted in Mehl, 189. See also William R. Shea, *The Magic of Numbers and Motion: The Scientific Career of René Descartes*, 105.

62   *brought the letters in question to Kepler and made his acquaintance*: Shea, 105.

62   *aware of Kepler's work through his friend Beeckman*: Adam and Tannery (1986), X:23.

62   *published it in* Mysterium cosmographicum *(1596)*: The treatise was written in July 1595, and so Kepler, who was born on December 27, 1571, was twenty-three years old at the time of his discovery. See Mehl, 17 n. 9, for details.

CHAPTER 5: THE ATHENIANS ARE VEXED BY A PERSISTENT ANCIENT PLAGUE

Page

67   *at two different locations separated by a known distance*: This was one of the greatest scientific achievements in antiquity. Eratosthenes measured that sunlight made an angle of a fiftieth of a circle (about seven degrees) from the perpendicular in Alexandria, and cast no shadow (meaning the angle was zero) in Syene in Upper Egypt. The two cities are 500 miles apart (5,000 stadia). Hence he computed the circumference of the earth to be $50 \times 500 = 25,000$ miles (250,000 stadia). For more complete details see Pasachoff, *Astronomy*, 15.

67   *to help his countrymen rid themselves of the plague*: Heath, *A History of Greek Mathematics*, I:246–60.

67   *say, 1,000 cubic meters*: I am using meters in this example not because the meter is an ancient measure (it is not), but rather because it is the only unit that allows me to use nice whole numbers such as 1,000, to make the mathematical example simple and illustrative while maintaining a reasonable size for a temple.

CHAPTER 6: THE MEETING WITH FAULHABER AND THE BATTLE OF PRAGUE

Page

72   *the mystic-mathematician Johann Faulhaber*: Baillet (1691), I:67.

72   *Descartes' earlier biographer, Daniel Lipstorp*: Lipstorp, *Specimina philosophiae*

cartesianae, 78–79. The story is reprinted in Adam and Tannery (1986), X:252–53.

73   *his own book, the Géométrie, published in 1637:* See Kenneth L. Manders, "Descartes and Faulhaber," *Bulletin Cartésien: Archives de Philosophie* 58, cahier 3 (1995), 1–12.

73   *Mehl concluded that Faulhaber and Descartes were close friends:* Mehl, 193.

74   *why Descartes chose this particular pseudonym:* Kurt Hawlitschek, "Die Deutschlandreise des René Descartes," *Berichte zur Wissenschaftsgechichte* 25 (2002), 240.

74   *was born in Ulm and was trained as a weaver:* Kurt Hawlitschek, *Johann Faulhaber 1580–1635: Eine Blütezeit der mathematischen Wissenschaften in Ulm*, 13.

76   *the meeting between Descartes and Faulhaber:* Baillet (1691), I:68.

76   *"I want you to enter my study":* Baillet (1691), I:68.

76   *"Cubic Cossic Pleasure Garden of All Sorts of Beautiful Algebraic Examples":* Lipstorp, *Specimina philosophiae cartesianae*, 79. I am indebted to Dr. Kurt Hawlitschek of Ulm for bringing this quotation to my attention.

77   *problems in Roth's book, and solved them as well:* Baillet (1691), I:69.

78   *the actual fighting since he was a volunteer:* Baillet (1691), I:73.

80   *as he had hoped to do two years earlier:* Adam and Tannery (1986), X:22.

CHAPTER 7: THE BROTHERHOOD

Page

83   *Baillet tells us:* Baillet (1692), 29.

83   *associated with the Rosicrucian order—Johann Faulhaber:* Richard Watson, *Cogito, Ergo Sum: The Life of René Descartes*, 103, argues that Descartes had already met a Rosicrucian before his trip to Germany. According to Watson, Descartes' friend in Holland Cornelius van Hogeland was a Rosicrucian.

86   *"After six score years, I shall be found":* Baillet (1691), I:89.

86   *The brothers made the following six rules:* Baillet (1691), I:90.

87   *"like news of a Second Coming":* Baillet (1691), I:92.

87   *the flaming star is passed around:* Anonymous, *Chevalier de l' Aigle du Pelican ou Rosecroix*, 5–7.

88   *Pythagorean theorem and early ideas about irrational numbers:* Watson argues in *Cogito, Ergo Sum*, 103–4, that scholars who persist in claiming that the Rosicrucians never existed lack an understanding of the nature of secret societies and how they operate.

89   *arrested by the Jesuits shortly after the publication appeared:* See Andreä, *Adam Haslmayr, der erste Verkünder der Manifeste de Rosenkreuzer,* 20.

CHAPTER 8: SWORDS AT SEA AND A MEETING IN THE MARAIS
Page

93   *the effects of the hostilities on the inhabitants of this region:* Baillet (1691), I;101.

93   *"could have been fatal for him":* Baillet (1691), I:101.

95   *"conduct him to his destination as peacefully as possible":* Baillet (1692), 49.

97   *"lodged somewhere in the Marais of the Temple in Paris":* Visitors to the Marais today can see both the rue du Temple and the rue Vielle du Temple.

97   *"in a manner imperceptible to the senses":* Baillet (1692), 55.

99   *in the form of mutual attacks:* Information on Marin Mersenne comes mostly from Vittorio Boria, "Marin Mersenne: Educator of Scientists."

100   *the worldwide correspondence he received and sent:* Boria, 91.

CHAPTER 9: DESCARTES AND THE ROSICRUCIANS
Page

102   *by members of the Brotherhood of the Rosy Cross:* Mehl, 31–36.

102   *used the term "Olympic" to mean* intelligible or comprehensible: Mehl, 31.

102   *code words for philosophy, magic, and alchemy:* Mehl, 32.

102   *described by Oswald Croll in his* Basilica chymica: Descartes to Mersenne, February 9, 1639, in Adam and Tannery (1988), II:498.

103   *eventually led to the decline of the order:* Yates, *The Rosicrucian Enlightenment,* 15–29.

103   *three years before the publication of this text:* Mehl, 37.

105   *as well as about their orbits in the sky:* Mehl, 43–45.

105   *The Rosicrucians named their principle* Est, Non est: Mehl, 104–6.

106   *selling his compass for use in engineering and for military purposes:* Boyer and Merzbach, *History of Mathematics,* 320.

106   *A published description of Faulhaber's qualifications included the following:* Testimonial about Johann Faulhaber by Johann Remmelin (1583–1632), published in Ulm in 1620, translated from the German by Kenneth L. Manders, in "Descartes and Faulhaber," *Bulletin Cartésien: Archives de Philosophie* 58, cahier 3 (1995), 2.

107   *Jacques Maritain says the following:* Maritain, *The Dream of Descartes,* 18.

108   *Hawlitschek hypothesizes in his book:* Hawlitschek, *Johann Faulhaber 1580–1635: Eine Blütezeit der mathematischen Wissenschaften in Ulm.*

108   *meet Faulhaber so they could discuss mathematics:* Hawlitschek, "Die Deutschlandreise des René Descartes," *Berichte zur Wissenschaftsgechichte* 25 (2002), 235–38.

108   *which led him to invent his own, related devices:* Mehl, 43.

109   *similarities in the content of the two manuscripts were discovered:* Mehl, 194.

109   *"worthy of making their acquaintance":* Mehl, 212 n. 87.

110   *the first one appearing in mid-October:* See Ivo Schneider, *Johannes Faulhaber 1580–1635,* 18–19, which includes the actual astronomical table prepared by Faulhaber. The story also appears in Shea, *The Magic of Numbers,* 104, although the date is given as September 1 (which would be true according to the Julian, rather than Gregorian, calendar).

111   *"cabalistic log-arithmo-geometro-mantica":* Mehl, 207.

111   *in a very personal, insulting way:* Mehl, 214.

CHAPTER 10: ITALIAN CREATIONS

Page

114   *"at least I may become more capable":* Baillet (1692), 56.

115   *the victory and the battle never took place, and were pure fiction:* Frederic C. Lane, *Venice: A Maritime Republic,* 57.

116   *as we would obtain today by solving the equation* $x+(1/7)x=19$: Boyer and Merzbach, 15–16.

117   *giving the two roots, or solutions, of this equation:* $x_{1,2}=\dfrac{-b\pm\sqrt{b^2-4ac}}{2a}$

118   *people had been trying for many centuries:* Boyer and Merzbach, 283.

118   *and sometimes a professorship at a university:* Jean Pierre Escofier, *Galois Theory,* 14.

118   *the coefficient of* x^3 *is 1, and there is no* x^2 *term:* Escofier, 14.

119   *a poem in Italian in which he embedded his formula:* Escofier, 14.

120–21 *work done a century earlier by the Italians:* The formulas developed by Tartaglia and Cardano and Ferrari are very complicated. Here is one of them, a general formula for the solution of the cubic equation $x^3+qx-r=0$. The solution is
$$\sqrt[3]{\frac{r}{2}+\sqrt{\frac{r^2}{4}+\frac{q^3}{27}}}+\sqrt[3]{\frac{r}{2}-\sqrt{\frac{r^2}{4}+\frac{q^3}{27}}}$$
in which the cube roots are varied so that their product is always $-q/3$.

121   *revive the use of the equal sign we use today:* Florian Cajori, *A History of Mathematical Notations,* I:300.

Notes

CHAPTER 11: A DUEL AT ORLÉANS, AND THE
SIEGE OF LA ROCHELLE

Page

123   *a wife "of good birth and much merit"*: Baillet (1691), II:501.

124   *remained known only as "Father P."*: Baillet (1691), II:501.

125   *Baillet tells us*: Baillet (1692), 69.

127   *who wanted to observe the siege*: Baillet (1691), I:155.

127   *"well appreciated by Cardinal Richelieu"*: Baillet (1691), I:157.

128   *was there, meeting the British officers*: Baillet (1691), I:159.

128   *tried to eat the leather of belts and boots*: Information about the siege obtained
      from documents kept at the Protestant Museum of La Rochelle.

129   *troops with their guns and ample ammunition*: Today, La Rochelle is a popular
      tourist destination and a thriving resort town. The old part of the city, dating
      from the twelfth to sixteenth centuries, looks much as it did in centuries past,
      and the medieval walls of the harbor with their towers still stand today. But
      the people of La Rochelle, most of whom are Catholic, as is the case through-
      out France, clearly are not proud of their history. Finding any reminders of the
      great siege of 1628 is very difficult, and the tourist office in town can supply
      visitors with no brochures or any kind of information whatsoever about this
      part of the city's history.

CHAPTER 12: THE MOVE TO HOLLAND AND THE
GHOST OF GALILEO

Page

132   *active population enjoyed the fruits of peace*: Descartes, *Discours de la Méthode*,
      edited by Etienne Gilson.

132   *had also contributed to his decision to leave France*: Gustave Cohen, *Ecrivains
      français en Hollande dans la première moitié du XVIIe siècle*, 402–9.

133   *"by all rights declare it as your own"*: Quoted in Jean-Marie Beyssade, *Etudes
      sur Descartes*, 33.

134   *"to make him ashamed, especially if I had his letter"*: Descartes to Mersenne in
      1629, quoted in Stephen Gaukroger, *Descartes: An Intellectual Biography*, 223.

135   *"had learned many things from you"*: Quoted in Varaut, 109.

135   *"which you describe by the name mathematico-physics"*: Quoted in Michel
      Fichant, *Science et métaphysique dans Descartes et Leibniz*, 19.

136   *"of ants and small worms"*: Quoted in Beyssade, *Etudes*, 33.

136   *In October 1629, Descartes started to work on a book:* For this date see Gilson's edition of *Discours de la Méthode,* 103 n. 3.

137   *as he later described his resolution:* Beyssade, *Etudes,* 36.

137   *that could enrage the powerful Inquisition:* Beyssade. *Etudes,* 40.

138   *fifth part of his Discourse, Descartes wrote the following:* Descartes, *Discours de la Méthode,* Gallimard ed., 111, author's translation.

139   *An excerpt follows:* Descartes to Mersenne, April 15, 1630, in Adam and Tannery (1974), I:145, author's translation.

140   *within his theory of the universe:* Fichant, 22.

140   *deciphering whether a symbol was a number or an abstract sign:* Fichant, 26.

141   *Hiding his physics by way of a "fable" was one more layer:* J. P. Cavaillé, *Descartes: La fable du monde,* 1.

143   *The letter is datelined Deventer, end of February 1634:* F. Alquié (1997), 492–93.

CHAPTER 13: A SECRET AFFAIR
Page

145   *a pretty servant named Hélène Jans:* Some scholars believe that Jans was the name of her father.

146   *"ten years now that God has removed me from that dangerous engagement":* Varaut, 139.

146   *"prescribes for those who live in bachelorhood":* Baillet (1691), II:89.

146   *perhaps to work as a maid for his landlady:* Gaukroger, *Descartes* (1995), p. 333.

148   *other mathematicians of the day:* Varaut, 141.

CHAPTER 14: DESCARTES' PHILOSOPHY AND THE
DISCOURSE ON THE METHOD
Page

151   *impervious to the dangers of skepticism:* F. Copleston, *A History of Philosophy,* IV:66–67.

152   *never accepting the authority of any previous philosophy:* F. Copleston, 67.

152   *The Discourse was Descartes' first published book:* F. de Buzon, preface, in Descartes, *Discours de la Méthode,* Gallimard ed., 7.

153   *one that has no center and whose dimensions are infinite:* F. de Buzon in Descartes, *Discours,* Gallimard ed., 9.

154   *sanitized scientific writings, and published them*: F. de Buzon in Descartes, *Discours*, Gallimard ed., 11.

154   *when it was withdrawn from publication*: F. de Buzon in Descartes, *Discours*, Gallimard ed., 11.

154   *"because of its certitude and its reasoning"*: Descartes, *Discours de la Méthode*, edited by Etienne Gilson, 52.

155   *"march forward with confidence in this life"*: Descartes, *Discours*, edited by Etienne Gilson, 56.

156   *"rolling here and there in the world"*: Descartes, *Discours*, edited by Etienne Gilson, 84.

159   *alluding to the association with the brotherhood*: Mehl, 87.

159   *"knowing more than they know"*: Descartes, *Discours*, Gallimard ed., 82, author's translation.

159   *"one would consider the most curious"*: Descartes, *Discours*, Gallimard ed., 78, author's translation.

160   *dealing with special knowledge: magic, astrology, and alchemy*: F. de Buzon in Descartes, *Discours*, edited by Gallimard, 78 n. 2. See also Descartes, *Discours*, Etienne Gilson, 49 n. 2.

160   *the problem solved in his secret notebook*: F. de Buzon in Descartes, *Discours*, Gallimard ed., 93 n. 1.

160   *"sum of all the science of pure mathematics"*: Fichant, 24.

CHAPTER 15: DESCARTES UNDERSTANDS THE ANCIENT DELIAN MYSTERY

Page

164   *on the second page of his Géométrie*: Adam and Tannery (1982), VI:370.

165   *neither are any other higher-order roots*: See the upcoming book by Simon Winchester, *Fatal Equation* (HarperCollins, 2009), for the story of the life of Galois.

CHAPTER 16: PRINCESS ELIZABETH

Page

168   *set his dogs on the impudent peasant*: J.-M. Beyssade and M. Beyssade, eds., *Descartes: Correspondence avec Elizabeth*, 24.

169   *"mysteries of nature as well as geometry"*: Baillet (1691), II:233.

171   *"both disciplines are equally easy to understand"*: Baillet (1691), II:233.

172   *to take care of a fellow royal in distress*: Gaukroger, *Descartes*, 385.

173   *wanted to devote her life to studying it*: Baillet (1691), II:231.

CHAPTER 17: THE INTRIGUES OF UTRECHT
Page

175   *"only because we clearly and distinctly perceive this"*: Adam and Tannery (1983), VII:214. See also Gaukroger, *Descartes*, 343.

177   *"belongs to the Society of the Brothers of the Rosy Cross"*: Quoted in Mehl, 92.

179   *helping him promote his teachings in Holland*: Varaut, 235.

CHAPTER 18: THE CALL OF THE QUEEN
Page

181   *"the most intimate secrets of his heart"*: Baillet (1691), II:242.

182   *"ever love me because I resemble you in any way"*: Baillet (1691), II:243.

182   *Descartes wrote the following curious passage*: Descartes to Chanut, from Egmond, Holland, November 1, 1646, in Beyssade and Beyssade, 245–46.

186   *"the letter Your Majesty has written me"*: Beyssade and Beyssade, 284.

186   *Alexander the Great comes to mind*: Varaut, 254.

187   *"place where I might do better"*: Beyssade and Beyssade, 281.

188   *"from a person of another religion"*: Varaut, 258.

191   *the last letter Descartes would write to her*: Descartes to Elizabeth, from Stockholm, October 9, 1649, in Beyssade and Beyssade, 234–35.

192   *"My desire to return to my desert grows every day more and more"*: Varaut, 269.

CHAPTER 19: THE MYSTERIOUS DEATH OF DESCARTES
Page

194   *"theologians of Utrecht and Leyden had declared upon him"*: Baillet (1691), II:417.

195   *to offer his services to the ailing philosopher*: Baillet (1691), II:417.

195   *"as an adult in good health without bleeding"*: Reported in Baillet (1692), 49.

196   *"I will die with more contentment if I do not see him"*: Baillet (1691), II:418.

197   *did not take well to his philosophy*: Baillet (1691), II:415.

197   *the claim that Descartes was poisoned:* Varaut, 271–81.

198   *the man she called "My Illustrious Master":* Baillet (1692), 268.

199   *paying for the expenses of Descartes' funeral:* Baillet (1691), II:425.

199   *in the Church of Sainte-Geneviève-du-Mont in Paris:* Baillet (1692), 270.

199   *in the ancient Church of Saint-Germain-des-Prés:* These details are from Geneviève Rodis-Lewis, in John Cottingham, ed., *The Cambridge Companion to Descartes,* 57 n. 74.

200   *be buried with the rest of his bones:* See Berzelius to Cuvier, April 6, 1821, in Adam and Tannery (1983), XII:618–19.

200   *the Musée de l'Homme (the Museum of Man) in Paris:* The museum is at the Trocadéro, across from the Eiffel Tower, in Paris. I tried to decipher the writings on the skull when I visited the museum in the summer of 2004. I could make out only the word "Stockholm" and a date that looks like 1660 or 1666.

201   *the deceased philosopher's valet, Henry Schluter:* Adam and Tannery (1986), X:1, after Baillet.

201   *making a small fortune on these items a few years later:* Adam and Tannery (1986), X:1, after Baillet.

202   *Chanut took all of these items under his "particular protection":* Adam and Tannery (1986), X:3, after Baillet.

203   *neither time nor patience for publication requests:* Adam and Tannery (1974), I:xvii (based on Lipstorp).

205   *"the Catholic, Apostolic, and Roman religion":* Baillet (1691), II:432.

CHAPTER 20: LEIBNIZ'S QUEST FOR DESCARTES' SECRET
Page

207   *and np lead back to y:* G. W. Leibniz, *Recherches générales sur l'analyse des notions et des vérités,* 136.

208   *books that were above his level:* E. J. Aiton, *Leibniz: A Biography,* 12.

208   *Leibniz also studied Bacon, Hobbes, Galileo, and Descartes:* Jean-Michel Robert, *Leibniz, vie et oeuvre,* 11.

208   *only through work in mathematics later in life:* Bertrand Russell, *The Philosophy of Leibniz,* 6.

208   *stamping on it his own unique impression:* Marc Parmentier, *La naissance du calcul différentiel,* 15.

208   *in jeopardy of losing his academic position:* W. Hestermeyer, *Paedagogia mathematica,* 51.

209   *"can do away with the flaws in the Cartesian doubt"*: Paul Schrecker, ed., G. W. *Leibniz: Opuscules philosophiques choisis*, 31.

209   *and the Latin manuscript* Cartesii opera philosophica: Yvon Belaval, *Leibniz critique de Descartes*, 9.

210   *offered Leibniz admittance to the society*: Aiton, 24.

211   *he would be paid what he was owed*: Aiton, 37.

213   *Latin name for Germany, making it read*: F. Alquié, ed., *Descartes: Oeuvres philosophiques*, I:45. Alquié hypothesizes that there is possibility that it was Foucher de Careil who added "Germania" as an explanation of G when he handled Leibniz's copy of Descartes during his visit to the Hanover archive.

213   *discussed its finer points at length in letters*: Aiton, 84.

213   *taken right out of the* Fama fraternitatis: Yates, *The Rosicrucian Enlightenment*, 154.

213   *the alchemical society of Nuremberg*: Jean-Michel Robert, *Leibniz, vie et oeuvre*, 14.

214   *Clerselier had imposed very strict rules on the copying*: Costabel, *René Descartes*, viii.

218   *found and dated to several centuries before Plato*: See the historical note by Sir Thomas L. Heath in Euclid, *The Thirteen Books of the Elements*, 3:438.

220   *This made his work even more flawed*: Adam and Tannery (1986), X:259.

CHAPTER 21: LEIBNIZ BREAKS DESCARTES' CODE AND SOLVES THE MYSTERY

Page

223   *the spheres containing the Platonic solids and the planets*: See Pasachoff, *Astronomy*, 27.

225   *the page he was looking at stood for the five regular solids*: The two formulas from the notebook, appearing just above the two sequences of numbers, are mathematical tools for generating the sequences.

227   *three-dimensional polyhedron—regular or not*: It fails for a Möbius strip, which is a "pathological" three-dimensional object.

229   *banned the teaching of Cartesian philosophy in France*: Gaukroger, *Descartes*, 3.

230   *rendered him by "Monsieur Leibniz, a German mathematician"*: Baillet (1691), I:xxvi.

231   *"by which he could draw attention to himself"*: Leibniz, *Philosophical Papers and Letters*, 223.

Notes

231   *both Cartesian and anti-Cartesian elements at the same time:* Yvon Belaval, *Leibniz critique de Descartes,* 12.

232   *"his metaphysics is all these things":* Leibniz to Nicolas Malebranche, Hanover, January 23, 1679, in Leibniz, *Philosophical Papers and Letters,* 209.

233   *"nothing but deductions from Descartes":* Aiton, 56.

233   *"only a continuation and elaboration of Descartes' ideas":* Leibniz, *Stämliche Schriften und Briefe,* III.1:504–16.

234   *English mathematicians with whom he had ties:* Aiton, 65.

234   *Newton read books about Descartes' mathematics:* Boyer and Merzbach, 391.

235   *acknowledging the contributions of Galileo, Kepler, and Descartes:* E. T. Bell, *Men of Mathematics,* 93.

235   *she suffered a shock and died:* Carr, *Leibniz,* 9.

A TWENTY-FIRST-CENTURY EPILOGUE
Page

237   *on the history of our understanding of the solar system:* The observations were made by an international team of astronomers led by Professor John Seiradakis, the director of the Observatory of the Aristotle University of Thessaloniki.

239   *article in the* Notices of the American Mathematical Society: Jeffrey Weeks, "The Poincaré Dodecahedral Space and the Mystery of the Missing Fluctuations," *Notices of the American Mathematical Society,* June/July 2004, 610–19.

Adam, Charles. *Vie et oeuvres de Descartes*. Paris: Cerf, 1910.

Adam, Charles, and Paul Tannery, eds. *Oeuvres de Descartes*. 12 vols. Paris: Leopold Cerf, 1902–12.

———, eds. *Oeuvres de Descartes*. Rev. ed. 12 vols. Paris: Vrin, 1974–88.

Aiton, E. J. *Leibniz: A Biography*. Bristol, U.K.: Adam Hilger, 1985.

Alquié, Ferdinand. *La découverte métaphysique de l'homme chez Descartes*. Paris: Presses Universitaires de France, 1950.

———. *Descartes: L'homme et l'oeuvre*. Paris: Hatier-Boivin, 1956.

———, ed. *Descartes: Oeuvres philosophiques*. Vol. I, 1618–37. Paris: Garnier, 1997.

Andreä, Johann Valentin. *Adam Haslmayr, der erste Verkünder der Manifeste de Rosenkreuzer*. Amsterdam: Plikaan, 1994.

Anonymous (attributed to Johann Valentin Andreä). *Fama fraternitatis*. Cassel, 1614. Reprint, Stuttgart: Calwer Verlag, 1981.

Anonymous. *Chevalier de l'aigle du pelican ou Rosecroix*. Reprint, Nimes: Lacour, 1998.

Ariew, Roger, and Marjorie Grene, eds. *Descartes and His Contemporaries*. Chicago: University of Chicago Press, 1995.

Arnold, P. *Histoire des Rose-Croix et les origines de la Franc-Maçonnerie*. Paris: Mercure de France, 1990.

Baillet, Adrien. *La Vie de Monsieur Des-Cartes*. 2 vols. Paris: Daniel Horthemels (rue Saint-Jacques), 1691.

———. *Vie de Monsieur Descartes*. Abridged ed., 1692. Reprint, Paris: La Table Ronde, 1972.

Baker, Gordon, and Katherine J. Morris. *Descartes' Dualism*. New York: Routledge, 1996.

Beck, Leslie John. *The Method of Descartes*. Oxford, U.K.: Oxford University Press, 1952.

Beeckman, Isaac. *Journal 1604–1634*. Introduction and notes by C. de Waard. 4 vols. La Haye: M. Nijhoff, 1939–53.

Belaval, Y. *Leibniz critique de Descartes*. Paris: Gallimard, 1960.

Bell, E. T. *Men of Mathematics*. New York: Simon & Schuster, 1937.

Beyssade, Jean-Marie. *Etudes sur Descartes: L'histoire d'un esprit*. Paris: Editions du Seuil, 2001.

———. *La philosophie première de Descartes*. Paris: Flammarion, 1979.

Beyssade, Jean-Marie, and M. Beyssade, eds., *René Descartes: Correspondance avec Elizabeth et autres lettres*. Paris: Flammarion, 1989.

Bitbol-Hespériès, Annie. *Le principe de vie chez Descartes*. Paris: Vrin, 1990.

Blanchet, L. *Les antécédents historiques du "Je pense donc je suis."* Paris: F. Alcan, 1920.

Bloom, John J. *Descartes: His Moral Philosophy and Psychology*. New York: New York University Press, 1978.

Boria, Vittorio. "Marin Mersenne: Educator of Scientists." Doctoral dissertation, American University, 1989.

Boyer, Carl B. *A History of Mathematics*. New York: Wiley, 1968.

Boyer, Carl B., and Uta C. Merzbach. *A History of Mathematics*. 2nd edition. New York: Wiley, 1991.

Brahe, Tycho. *Opera omnia*. 15 vols. Edited by J. L. E. Dreyer. 1913–26.

Cahné, Pierre-Alain. *Un autre Descartes—le philosophe et son langage*. Paris: Vrin, 1980.

Cajori, Florian. *A History of Mathematical Notations*. Vols. I and II. New York: Dover, 1993.

Carr, Herbert W. *Leibniz*. New York: Dover, 1960.

Cavaillé, J.-P. *Descartes: La fable du monde*. Paris: Vrin, 1991.

Cohen, Gustave. *Ecrivains français en Hollande dans la première moitié du XVIIe siècle*. Paris: E. Champion, 1920.

Cole, John R. *The Olympian Dreams and Youthful Rebellion of René Descartes*. Chicago: University of Illinois Press, 1992.

Copleston, Frederick. *A History of Philosophy*. Vol. IV, *Modern Philosophy: From Descartes to Leibniz*. New York: Doubleday, 1994.

Costabel, Pierre. *Démarches originales de Descartes savant*. Paris: Vrin, 1982.

————, ed. *René Descartes: Exercises pour les éléments des solides*. Paris: Presses Universitaires de France, Epiméthée, 1987.

Cottingham, John. *Descartes*. New York: Oxford University Press, 1998.

————, ed. *The Cambridge Companion to Descartes*. New York: Cambridge University Press, 1992.

Cousin, Victor. *Histoire générale de la philosophie depuis les temps les plus anciens*. Paris, 1884.

Couturat, Louis. *La logique de Leibniz*. New York: George Olms Verlag, 1985.

Croll, Oswald. *Basilica chymica*. Frankfurt: G. Tampach, 1620.

Curley, Edwin. *Descartes Against the Skeptics*. Cambridge: Harvard University Press, 1978.

De Sacy, Samuel S. *Decartes*. Paris: Seuil, 1996.

————. *Descartes par lui-même*. Paris: Seuil, 1956.

Denissoff, Elie. *Descartes: Premier théoricien de la physique mathématique*. Paris: Louvain, 1970.

Descartes, René. *Discours de la méthode*. 1637. Republished, with notes by F. de Buzon, Paris: Gallimard, 1997.

————. *Discours de la méthode*. 1637. Republished and edited, with notes, by Etienne Gilson, Paris: Vrin, 1999.

————. *Discourse on the Method and Meditations on First Philosophy*. Edited by David Weissman. New Haven: Yale University Press, 1996.

————. *Philosophical Letters*. Translated and edited by Anthony Kenny. Oxford, U.K.: Clarendon Press, 1970.

————. *The Philosophical Works of Descartes*. Translated and edited by Elizabeth S. Haldane and G. R. T. Ross. Vols. I and II. Cambridge, U.K.: Cambridge University Press, 1968.

Dunn, Richard S. *The Age of Religious Wars*. New York: Norton, 1970.

Escofier, Jean Pierre. *Galois Theory*. New York: Springer-Verlag, 2001.

Euclid. *The Thirteen Books of the Elements*. Vol. 3; Books X–XIII. New York: Dover, 1956.

Faulhaber, Johann. *Arithmetischer Cubicossicher Lustgarten*. Tübingen, Germany: E. Cellius, 1604.

————. *Miracula arithmetica*. Augsburg: D. Francken, 1622.

————. *Numerus figuratus sive arithmetica arte mirabili inaudita nova constans*. Ulm, Germany, 1614

Fichant, Michel. *Science et métaphysique dans Descartes et Leibniz*. Paris: Presses Universitaires de France, 1998.

Foucher de Careil, Louis-Alexandre. *Oeuvres inédites de Descartes*. Paris: Foucher de Careil, 1860.

Gäbe, Lüder. *Descartes Selbskritik: Untersuchungen zur Philosophie des jungen Descartes*. Hamburg: Meiner, 1972.

Garber, Daniel. *Descartes' Metaphysical Physics*. Chicago: University of Chicago Press, 1992.

———. *Descartes Embodied: Reading Cartesian Philosophy Through Cartesian Science*. New York: Cambridge University Press, 2001.

Gaukroger, Stephen. *Cartesian Logic*. Oxford, U.K.: Oxford University Press, 1989.

———. *Descartes: An Intellectual Biography*. Oxford, U.K.: Clarendon Press, 1995.

Gilder, J., and A.-L. Gilder. *Heavenly Intrigue*. New York: Doubleday, 2004.

Gilson, E. *Index scholastico-cartésien*. New York: Burt Franklin, 1912.

Gouhier, Henri. *Essais sur Descartes*. Paris: Vrin, 1949.

———. *La pensée métaphysique de Descartes*. Paris: Vrin, 1969.

———. *Les premières pensées de Descartes*. Paris: Vrin, 1958.

Grimaldi, Nicolas. *L'expérience de la pensée dans la philosophie de Descartes*. Paris: Vrin, 1978.

Guenancia, Pierre. *Descartes*. Paris: Bordas, 1989.

Gueroult, Martial. *Descartes selon l'ordre des raisons*. Paris: Aubiers, 1953.

Hamelin, O. *Le système de Descartes*. Paris: Alcan, 1911.

Hawlitschek, Kurt. *Johann Faulhaber 1580–1635: Eine Blütezeit der mathematischen Wissenschaften in Ulm*. Ulm, Germany: Stadtbibliothek Ulm, 1995.

Heath, Sir Thomas. *A History of Greek Mathematics*. Vol. I. New York: Dover, 1981.

Hebenstreit, Johann-Baptist. *Cometen Fragstück aus der reinen Philosophica*. Ulm, Germany: J. Meder, 1618.

———. *De Cabbala log-arithmo-geometro-mantica*. Ulm, Germany: J. Meder, 1619.

Heindel, Max, and Augusta Heindel. *Histoire des Rose-Croix*. Lodève: Beaux Arts, 1988.

Hestermeyer, W. *Paedagogia mathematica*. Paderborn: Schöningh, 1969.

Holton, Gerald. *Thematic Origins of Scientific Thought*. Cambridge: Harvard University Press, 1973.

Hooker, Michael, ed. *Descartes: Critical and Interpretative Essays*. Baltimore: Johns Hopkins University Press, 1978.

Judovitz, Dalia. *Subjectivity and Representation in Descartes: The Origins of Modernity*. New York: Cambridge University Press, 1988.

Jullien, Vincent. *Descartes: La Géométrie de 1637*. Paris: Presses Universitaires de France, 1996.

Kepler, Johann. *Harmonices mundi*. Linz: G. Tampachius, 1619.

———. *Kanones pueriles*. Ulm, 1620.

———. *Mysterium cosmographicum*. Tübingen, 1596. Translated as *The Secret of the Universe* by A. M. Duncan. New York: Abaris, 1981.

Lane, Frederic C. *Venice: A Maritime Republic*. Baltimore: Johns Hopkins University Press, 1973.

Laporte, Jean. *Le rationalisme de Descartes*. Paris: Presses Universitaires de France, 1945.

Lefèvre, R. *La vocation de Descartes*. Paris: Presses Universitaires de France, 1956.

Leibniz, Gottfried Wilhelm. *De arte combinatoria*. Leipzig, 1666.

———. *Philosophical Papers and Letters*. Edited by L. E. Lömker. 2nd ed. Dordrecht, Holland: D. Reidel, 1970.

———. *Recherches générales sur l'analyse des notions et des vérités*. Edited by Jean Baptist Rauzy. Paris: Presses Universitaires de France, 1998.

———. *Stämliche Schriften und Briefe*. Berlin: Akademie Verlag, 1923.

Lenoble, Robert. *Mersenne ou la naissance du mécanisme*. Paris: Vrin, 1943.

Lipstorp, Daniel. *Specimina philosophiae cartesianae*. Leyden, 1653.

Mahoney, Michael Sean. *The Mathematical Career of Pierre de Fermat*. Princeton: Princeton University Press, 1973.

Marion, Jean-Luc. *Sur l'ontologie grise de Descartes*. Paris: Vrin, 1975.

———. *Sur le prisme métaphysique de Descartes*. Paris: Presses Universitaires de France, 1986.

———. *Sur la théologie blanche de Descartes*. Paris: Presses Universitaires de France, 1991.

———. *Questions cartésiennes*. Paris: Presses Universitaires de France, 1996.

McIntosh, Christopher. *The Rosy Cross Unveiled: The Rise, Mythology, and Rituals of an Occult Order*. Northamptonshire, U.K.: Aquarian Press, 1980.

Mehl, Edouard. *Descartes en Allemagne*. Strasbourg: Presses Universitaires de Strasbourg, 2001.

Mersenne, Marin. *Quaestiones celeberrimae in Genesim*. Paris: S. Cramoisy, 1623.

Mesnard, P. *Essai sur la morale de Descartes*. Paris: Boivin, 1936.

Milhaud, G. *Descartes savant*. Paris: Alcan, 1921.

Millet, J. *Descartes, sa vie, ses travaux, ses découvertes avant 1637*. Paris: Clermont, 1867.

Mögling, Daniel (under the pseudonym T. Schweighardt). *Speculum sophicum rhodostauroticum universale*. S.L. 1618.

Moreau, Denis. *Deux cartésiens: La polémique entre Antoine Arnauld et Nicolas Malebranche*. Paris: Vrin, 1999.

Morris, John M., ed. *Descartes Dictionary*. New York: Philosophical Library, 1971.

Naudé, Gabriel. *Instruction à la France sur la vérité de l'histoire des Frères de la Rose-Croix*. Paris: Juliot, 1623.

Parmentier, Marc. *La naissance du calcul différentiel*. Paris, Vrin, 1989.

Pasachoff, Jay. *Astronomy: From the Earth to the Universe*. 5th ed. New York: Saunders, 1995.

Reichenbach, Hans. *The Philosophy of Space and Time*. New York: Dover, 1958.

Robert, Jean-Michel. *Leibniz, vie et oeuvre*. Paris: Univers Poche, 2003.

Rodis-Lewis, Geneviève. *Descartes*. Paris: Calmann-Lévy, 1995.

———. *Descartes: His Life and Thought*. (English translation of the 1995 book.) Translated by Jane Marie Todd. Ithaca, N.Y.: Cornell University Press, 1995.

———. *Descartes: Textes et débats*. Paris: Poche, 1984.

———. *Le développement de la pensée de Descartes*. Paris: Vrin, 1997.

———. *L'oeuvre de Descartes*. Paris: Vrin, 1971.

Roth, Peter. *Arithmetica philosophica*. Nuremberg: J. Lantzenberger, 1608.

Russell, Bertrand. *The Philosophy of Leibniz*. London: Gordon and Breach, 1908.

Scheiner, Christoph. *Oculus hoc est: Fundamentum opticum*. Innsbruck: D. Agricola, 1619.

Schneider, Ivo. *Johannes Faulhaber 1580–1635: Rechenmeister in einer Welt des Umbruchs*. Berlin: Birkhauser Verlag, 1993.

Schrecker, Paul, ed. *G. W. Leibniz: Opuscules philosophiques choisis*. Paris: Vrin, 2001.

Scribano, Emanuela. *Guida alla lettura della Meditazioni metafsiche di Descartes*. Rome: Laterza, 1997.

Sebba, Gregor. *Bibliographia Cartesiana: A Critical Guide to the Descartes Literature, 1800–1960*. La Haye, 1964.

Shea, William R. *The Magic of Numbers and Motion: The Scientific Career of René Descartes*. Canton, Mass.: Science History Publications, 1991.

Simon, G. *Kepler, astronome, astrologue*. Paris: Gallimard, 1992.

Sorrell, Tom. *Descartes: A Very Short Introduction*. New York: Oxford University Press, 2000.

Sriven, J. *Les années d'apprentissage de Descartes (1596–1618)*. Paris: Vrin, 1928.

Tannery, Paul, and Cornelius de Waard, eds. *Correspondance du Père Marin Mersenne*. 17 vols. Paris: Centre National de la Recherche Scientifique, 1963–88.

Szpiro, George. *Kepler's Conjecture*. New York: Wiley, 2003.

Van Peursen, C. A. *Leibniz*. London: Faber and Faber, 1969.

Varaut, Jean-Marc. *Descartes: Un cavalier français*. Paris: Plon, 2002.

Verbeek, Theo, ed. *René Descartes et Martin Schoock: La querelle d'Utrecht*. Paris: Les Impressions Nouvelles, 1988.

Vrooman, J. R. *René Descartes: A Biography*. New York: G. P. Putnam's Sons, 1970.

Vuillemin, Jules. *Mathématiques et métaphysique chez Descartes*. Paris: Presses Universitaires de France, 1987.

Watson, Richard. *Cogito, Ergo Sum: The Life of René Descartes*. Boston: David R. Godine, 2002.

Williams, Bernard. *Descartes: The Project of Pure Enquiry*. New York: Penguin, 1978.

Wilson, C. *Leibniz's Metaphysics: A Historical and Comparative Study*. Princeton, N.J.: Princeton University Press, 1989.

Wilson, Margaret. *Descartes*. London: Routledge, 1978.

Yates, Frances A. *The Art of Memory*. Chicago: University of Chicago Press, 1966.

———. *The Rosicrucian Enlightenment*. London: Routledge and Kegan Paul, 1972.

Zweig, Paul. *The Heresy of Self-Love*. Princeton, N.J.: Princeton University Press, 1980.

ILLUSTRATION CREDITS

INDEX

Index

272

© DEBRA GROSS ACZEL

ABOUT THE AUTHOR

Amir D. Aczel, Ph.D., is a mathematician and the author of twelve books, including the international bestseller *Fermat's Last Theorem*, which was nominated for a *Los Angeles Times* Book Prize and has been translated into nineteen languages. Aczel has appeared on over thirty television programs, including the *CBS Evening News* and ABC's *Nightline*, on CNN and CNBC, and on over a hundred radio programs, including NPR's *Weekend Edition* and *Talk of the Nation: Science Friday*. In 2004 Aczel was awarded a Guggenheim Fellowship. He is a visiting scholar in the history of science at Harvard.